A CASEBOOK FOR BUSINESS STATISTICS:

Laboratories for Decision Making

Norean Radke Sharpe
Division of Mathematics and Science
Babson College

Abdul Ali
Division of Marketing
Babson College

Mark Edward Potter
Division of Finance
Babson College

With Additional Cases by
Steven Eriksen
Division of Mathematics and Science
Babson College

JOHN WILEY & SONS, INC.
New York • Chichester • Weinheim • Brisbane • Singapore • Toronto

Cover Photo: © Jon Riley/Tony Stone Images, NYC.

To order books or for customer service call 1-800-CALL-WILEY (225-5945).

ISBN 0-471-38240-x

Printed in the United States of America

10 9 8 7 6 5 4 3 2 1

Printed and bound by Victor Graphics, Inc.

This book is dedicated to our former students who had the courage to use experimental work, provide constructive input, and work hard to improve their conceptual understanding of statistics.

PREFACE

Motivation for Laboratories for Decision Making

When I first began teaching statistics in the late 1980's, applications and data were secondary to the teaching of theory and techniques. The focus of most courses continued to be derivation and computation. When I accepted my current position at a business school, I was encouraged by the emphasis on real data, and by the use of case studies in most aspects of the business curriculum. In addition, most faculty were supportive of the inclusion of laboratory exercises and cases in the elementary statistics courses to enhance conceptual learning.

Now, a weekly or biweekly laboratory session is standard in many of our required statistics courses. These lab sessions, which were originally conducted in a separate room with networked desktop computers for every two students, now are conducted in our regularly scheduled classroom – also with network capabilities – with students using their own laptops. The advantages of these modern technical capabilities are diverse. In my view, the main benefit is that the students (and the faculty) now accept these interactive case-driven and computer-enabled learning sessions as inherent and intrinsic to the learning of statistics.

I have used published cases from a variety of sources over the past several years with varying degrees of success – primarily due to a mismatch between the level of sophistication of the case and the level of introductory statistics students; the lack of business applications in the case; and the lack of specific questions providing guidance for a student learning statistics for the first time. However, every time I used cases, it was clear that using *any* of these cases was better than *not* using cases.

Thus the need for the following was clear:

- Cases/Labs at the *introductory level*;
- Integrated cases/labs for *cross-functional curricula*;
- Data sets containing *recent and real data.*

A collaborative effort with faculty in other disciplines, who have a vested interest in students learning statistics, produced this set of integrated cases/labs. The encouraging verbal and written feedback from our students was so overwhelming, that we owe the inspiration and motivation for this book to our former students. These students had the courage to use experimental work, provide constructive input, and work hard to improve their ability to conceptualize and think critically about statistics.

Features and Objectives of Laboratories for Decision Making

Thanks to the efforts of statistics educators, such as David Moore, George Cobb, and Richard Schaeffer, teachers during the 1990's witnessed major changes in the landscape of statistics education. The reformed philosophy in teaching statistics was best summarized in a report of the joint ASA/MAA statistics curriculum focus group in *Heeding the Call for Change: Suggestions for Curricular Action* (MAA Notes, Volume 22), which offered three recommendations:

- Emphasize Statistical Thinking;
- More Data and Concepts;
- Foster Active Learning.

Thankfully, most instructors of statistics now agree that exposure to *real data* is a necessary part of any statistics curriculum. In addition, recent curricular reforms – both at my own business institution, and at large public universities – have recognized the importance of including case studies in all levels of statistics courses.

In addition to providing cases for introductory statistical topics, we use the perspective of recent curricular reforms in statistics education and focus on:

- Statistical *concepts,* as well as tools;
- Specific student questions for both *analysis* and *discussion;* and
- Discussion of *real objectives* and *implications* of analysis.

Our primary objective and premise for this casebook is to provide a set of cases for the sequence of topics typically taught in an Introductory Business Statistics Course. Thus the cases were written to build in difficulty and complexity sequentially as the student progresses through the casebook. The cases cover topics ranging from descriptive graphical and numerical analysis, to hypothesis testing and regression, to times series and forecasting. The data sets for the case studies are primarily obtained from actual projects conducted by finance, marketing, or management science faculty.

Learning Environment for Laboratories for Decision Making

We discovered as we were writing and piloting these cases, that students and faculty prefer a variety of case approaches – some more proscriptive than others. Therefore, our cases may include as many as three types of questions:

- Conceptual Questions – emphasizing understanding of fundamental concepts;
- Analysis Questions – requiring active data analysis;
- Discussion Questions – encouraging interpretation and discussion of results.

The intended use of these labs/cases is in a laboratory environment where the students have access to networked computers. The lectures themselves may be large with multiple lab sections, or the analysis questions may be completed outside of class independently. We expect the students to be novice statistics students with a motivation to use statistical tools in other disciplines. However, we assume the students have a knowledge of elementary calculus, as well as a knowledge of basic financial terms.

The data sets for each of the labs/cases in this book are available for downloading from the website located at www.wiley.com/college/sharpe. Each data set was created in Excel, but is easily transferred into a file in Minitab (or other statistical package).

Many of these cases have been used for the past few years in a sequence of undergraduate statistics courses, which includes: *Statistics & Probability*, *Quantitative Modeling*, and *Applied Statistics with Applications in Forecasting*, in addition to a first course in *Statistics* for our evening business students working toward an MBA. The final version of these labs/cases is the result of years of piloting and rewriting in response to student and faculty feedback.

We greatly appreciate the valuable insights provided by these students and other faculty during the early stages of the writing of this book. Please send future comments to the email address below and we hope your students enjoy the cases as much as ours have.

Norean Radke Sharpe (sharpen@babson.edu)

ACKNOWLEDGMENTS

First, we thank Professor Steven Eriksen at Babson College, who contributed cases to this casebook from his own vast consulting experiences. In addition to providing additional cases, Professor Eriksen also edited many of the cases, and provided valuable direction and commentary. His guidance provided both perspective and support.

Second, we thank the faculty at Babson College, who piloted many of these cases in their courses and contributed valuable feedback. This stage of the process was essential to the enhancement and final publication of the cases/labs.

Third, we thank our research assistant, Bethanie Lydon, who worked tirelessly in the research and editing of many of these cases.

Finally, we thank our reviewers who encouraged and enabled us to improve our work, and Debbie Berridge, editor at John Wiley & Sons, Inc., who believed in the project.

CURRICULUM IN CONTEXT: TEACHING WITH CASE STUDIES IN STATISTICS COURSES[1]

Introduction

The Harvard Business School has used case studies as its primary teaching tool ever since their first appearance in 1920 [3], and now produces between 600-700 cases (including field-based cases and "case-related" teaching materials) each year [18]. In the past decade, the success of using cases to teach has transcended the traditional business disciplines and case studies are now used to teach courses in statistics at a variety of levels. This increased interest in non-business oriented cases has generated several new casebooks for statistics that provide teaching alternatives, with a wide array of application areas at multiple levels of difficulty.

Although most business disciplines have been actively using and producing case studies as part of their standard curriculum, cases have not been considered an integral part of the undergraduate statistics curriculum. Thus we began experimenting with the case method several years ago in an effort to find pedagogical tools that could be used to discuss statistical concepts in a real context. After using cases now at all levels – introductory, intermediate, and advanced – in our undergraduate applied statistics sequence, we are now writing cases for our integrated curriculum (with finance and marketing). Cases motivate students to learn about the subject of statistics; challenge students to think critically and creatively about open-ended problems; and help students to visualize and apply statistical concepts in other disciplines. Here, I reinforce how the case study approach complements recent reforms in statistics education, such as using real data, motivating active learning, and improving written work.

Why Use Case Studies?

Cases Use Real Data In Context: Statisticians widely believe that statistics should be taught using real data (see, for example Bradstreet [2]; Cobb [6]; Lock [9]; Moore [10]; and Singer and Willett [17]). In a benchmark summary of statistics education reform, Cobb [6] recommends using "More Data and Concepts: Less Theory, Fewer Recipes." Cobb bases these recommendations on the belief that working with real data helps students learn statistical concepts and that students learn by constructing, doing, and reapplying [6]. From these assumptions about learning, it follows that students need to spend time constructing databases and applying statistical methods in real situations. The appropriate application of statistical procedures depends on the data, the context in which the data were collected, and the context in which the data are to be interpreted. Thus to fully appreciate both the power and limitations of data, data analyses need to be conducted in context.

[1] This paper by Norean Radke Sharpe is excerpted with permission from the recent Mathematical Association of America (MAA) Notes Volume entitled *Teaching Statistics: Resources for Undergraduate Instructors* edited by Thomas Moore. The original was based on a working paper prepared and submitted for publication in September 1997.

Cases require analysis and interpretation of data in context. Cases provide a description of a problem in industry that requires data analysis. The focus is on decision making and the application area of the problems is varied; the industries represented include financial, medical, pharmaceutical, manufacturing, and transportation. A description of the primary data source, data collection process (i.e., surveys, site visits, etc.), and variable definitions is included in the case along with the data. Cases vary in levels of complexity and stages of completion: 1) the "fully-analyzed" cases describe the appropriate paradigm for descriptive and inferential analysis in detail; 2) the "proscriptive" cases provide specific questions to guide the student through the analysis; and 3) the "open-ended" cases require the student to select a set of statistical procedures appropriate for analyzing the data and making a decision. All of these cases are useful to encourage an appreciation for the context (time frame, limitations, assumptions, etc.) of the data and teach the implications of choosing suitable tools in different decision making contexts.

Cases Motivate Student Involvement: Student learning is improved if students are actively involved in the learning process [14]. If students are actively working with information, rather than passively receiving it, they have a better chance of understanding it. Cobb [6] recommends that we "Foster Active Learning." More specifically, Garfield [7] suggests that students should involve themselves in the learning process, practice doing, receive feedback, and "confront their own misconceptions." Students need the opportunity to read about concepts, experiment with their interpretation of these concepts, raise questions about these interpretations, and participate in constructive interactive exchanges about their questions. Moore [12] describes a model that strives to achieve the goals of statistics reform was described as one in which students learn through activities, and teachers guide – as opposed to provide – student learning.

Cases motivate student preparation and active student involvement in data analysis and interpretation. To accomplish the goal of actively involving students in the analysis process, we use all three types of cases – fully-analyzed, proscriptive, and open-ended – in a sequence of increasing difficulty. First, students read and discuss good examples of data analysis required for consulting: method selection, output interpretation, and a decision-oriented recommendation. Second, students complete unfinished cases, optimally as part of an interactive laboratory component, and accompany their results with a written report. Third, students answer an open-ended question for a "client," given only the objective, the data, and background information on the data. To effectively complete this series of cases, students need to be an active analyst – determining, conducting, and interpreting the analysis in the context of the data.

Cases Encourage Oral & Written Communication: As the writing-across-the-curriculum movement was gaining momentum in the early nineties, statistics educators recognized the importance of strong oral and written communication skills for their students (see, for example Cobb [6]; Iversen [8]; Radke Sharpe [16]). A pedagogical approach that emphasizes these communication skills with statistical concepts is essential; a statistician must be prepared to convey statistical results and conclusions to clients in industry– including both practitioners and non-practitioners of statistics. In recent talks on this subject, Moore [11,12] has emphasized the importance of communication and cooperation in relationship to introducing students to "statistics in practice." If we assume that students are going to use statistics in an applied context, then the importance of communication is clear. Students need experience with communicating their own questions to instructors and other students; communicating their interpretations as a team member to other members of the team, and communicating their conclusions to an audience outside the statistical community. In a recent discourse on the impact of technology on statistics education, the themes of human interaction, facilitation, and feedback were defended as important aspects of the educational process [13].

Cases enrich the educational process by providing templates of good writing in the discipline and providing opportunities for oral discourse on statistical issues. The fully analyzed cases provide templates of approaches to writing about statistics. More specifically, these templates demonstrate appropriate use of tables and figures, mathematical expressions, and prose. They also provide examples of how to structure a report based on statistical analysis. When students are required to produce reports for the open-ended cases, their own documents improve over time after repeated revisions and exposure to other exemplary reports (obtained from both the casebook and other students). In addition to increasing emphasis on written work, cases stimulate class discussion. Because cases present the data in the context of a real issue in industry, most students find them intrinsically interesting, and sometimes even controversial. This increased level of interest among the students seems to generate enthusiasm for the class discussions, as well as for the statistical concepts themselves.

One Approach to Teaching Statistics with Cases

Just as with other teaching materials, different pedagogical approaches are possible with case studies, and several casebooks are now available. We chose to use a casebook in conjunction with a standard business statistics textbook [1]. I required my students to purchase both books and have not received negative feedback regarding the increased cost of purchasing two books. In fact, students have used their textbook as a "reference" book and their casebook as the primary reading material for the course (their casebooks were visibly used – marked up and worn). More recently, in an effort to provide students with a shorter and less expensive casebook, I have had success in working with a publisher to combine a few of my own cases with previously published cases and create a customized casebook (see [5]). Both of our introductory and intermediate courses are required of all undergraduates, and thus are offered multiple times every semester. The first course is an intensive one-semester survey of descriptive and inferential statistics that concludes with simple regression. The second course is a continuation of these topics, and includes multiple regression, time series, forecasting, quality control, simulation, and decision analysis. Our students typically have a wide range of academic ability, from second-year students who failed previous semesters of statistics to first-year students who placed out of calculus by taking the AP exam. Here, I attempt to summarize a few insights into teaching with cases. Since the case method has been growing in popularity, there are many academics with experience in this approach, who are able to provide guidance. New case instructors might consider attending workshops and reading the case literature (see for example, Parr and Smith [15], who provide a detailed template for developing a case-based course).

Assign cases in advance for preparation for class discussion: Requiring advance preparation raises expectation and performance. We expect students to read, understand, interpret, and explain the concepts they encounter in the cases. For fully-analyzed cases, we expect students to explain ***why*** a specific statistical technique was chosen, to interpret the results, and to agree or disagree with the conclusions of the author. By initially using cases that are fully analyzed, we provide students with a written model of the relationship between the data, the analysis, and the conclusions. We encourage them to explore their understanding of concepts, raise questions about these interpretations, and comment on other students' interpretations. The expectation of prior preparation of students increases their self-reliance and self-discipline, enhances teamwork skills among the students, and encourages students to become more responsible for their own learning. The focus of the responsibility for the learning experience is removed from the instructor and returned to the student. Moore [12] suggested this idea strongly when he said, " learning is a student responsibility." If the new paradigm of teacher as motivator, mentor, and guide is to be successful, then students will have to become more autonomous and active in the entire learning process.

Ask students to introduce cases, ask and answer questions about cases, and respond to other students' questions: Case discussions emphasize oral communication of statistical questions, ideas, and interpretations. This approach also encourages interaction with the instructor and other students. To create an encouraging environment for class discussion, we try to consistently provide every student an opportunity to comment on the case over the period of two case discussions. Probably more important from the student perspective, a reward structure for class participation is in effect on case discussion days. The dynamic of an active class discussion, as opposed to a discourse on the topic by the instructor, motivates the students to learn more about the application of statistics to other functional areas. The encouragement to participate also seems to raise questions that are less frequently asked, internally confusing, and more interesting. Frequent oral discussions in which the students are expected to participate enhance the interest level, intellect, and insight of the students in the class.

Ask students to produce statistical output on-line in class, bring prepared output to class, and/or complete cases as laboratory projects: As the students become comfortable with the case method, we expect them to prepare the open-ended cases, as if they were a consultant in industry. Students access the databases on their own outside of class and either bring their output to class, or are prepared to create it using the technology in the classroom. We also use cases as laboratory assignments followed by required written reports. A popular final project is one in which teams of students obtain and analyze a unique data set and write their own case study. We have even revised and rewritten some of these projects for use as later case studies. As with most teaching materials, instructors should customize the use of cases to their own course topics, classroom dynamics, class size, student academic abilities, and student communication skills. When students choose their own data sets, conduct their own analyses, and write their own case studies, they learn more about the practice of statistics, gain experience working in a team, and feel a greater sense of accomplishment.

Student Feedback

At the end of prior semesters, I have distributed a case evaluation survey and asked for student opinion on: (1) the effectiveness of the cases in helping to explain statistical concepts and techniques; (2) the strength of student recommendations to continue using cases in the course; and (3) the usefulness of the fully-analyzed cases. The results were extremely positive: approximately 80% of students responded that the cases were effective at explaining concepts and "techniques covered in class." More impressively, approximately 90% responded that they would "recommend ...the use...[of] cases again in this statistics course." When asked if they would recommend "the use of *more* cases in the statistics course," student responses were mixed. My interpretation of these mixed responses is that our introductory statistics course is densely packed with material – if more cases are added, something else must be subtracted. (Many institutions may have similarly packed courses and share the dilemma of how to experiment with pedagogical innovations, while still covering the required set of topics – breadth vs. depth.) Finally, on the question about the usefulness of fully-analyzed cases, approximately 90% found these cases to be "helpful."

From the questions on the survey that required a written response, I have concluded that the most effective and most popular cases were the cases on regression. I should also note that some students wrote that they would *not* have understood a particular statistical concept (e.g., the Central Limit Theorem) without the associated discussion motivated by the case. Over the past few years, several comments on cases have appeared on my standard college student opinion surveys. No negative remarks about cases have appeared on any of my opinion surveys and only one individual asked for fewer cases, because of the "fast-pace" of the course. Below is a sample of these student comments:

> *"The cases were good at explaining the information in the class."*
>
> *"[Use] more case studies – that's how …everyone learned the material."*
>
> *"[Use] more cases with detailed, maybe two day discussions, and with homework questions."*
>
> *"Cases were excellent in helping me understand the course. Maybe the whole class can turn into a case-based class."*
>
> *"The use of cases was very helpful, especially since it showed how the statistical material was applied in real situations. I think cases should be used more frequently."*
>
> *"The cases helped a lot and gave an interesting twist to the course."*
>
> *"Very relevant to the real world – I liked the cases and the review of the cases was very helpful in understanding the material."*

Conclusion

In summary, cases facilitate recent reforms in statistics education, such as using real data, motivating student involvement, and providing examples and opportunities for written work. In addition, cases encourage class discussions and motivate the use of databases for a laboratory component of a course. Finally, student feedback suggests that cases are effective at assisting students in the explanation of statistical concepts and techniques. As a result of the training cases provide in decision-making, students, faculty, practitioners, and employers in industry should all benefit.

Acknowledgment

The author would like to acknowledge George Cobb for his thoughtful and insightful comments on an earlier version of this manuscript.

References

[1] Berenson, M.L. and Levine, D.M. (1996), *Basic Business Statistics* (Sixth Edition), Englewood Cliffs, NJ: Prentice Hall.

[2] Bradstreet, T.E. (1996), "Teaching introductory statistics courses so that nonstatisticians experience statistical reasoning," *The American Statistician*, 50 (1).

[3] Budman, M. (1995), "The Business of Cases," *Across The Board,* 32(2).

[4] Chatterjee, S., Handcock, M.S., and Simonoff, J.S. (1995), *A Casebook for a First Course in Statistics and Data Analysis,* New York: John Wiley & Sons.

[5] Chatterjee, S., Handcock, M.S., and Simonoff, J.S. (selected cases) with additional cases by Radke Sharpe, N. (1997), *Casebook for Statistics and Data Analysis,* New York: John Wiley & Sons.

[6] Cobb, G. (1992) "Teaching Statistics," in L.A. Steen (ed.), *Heeding the Call for Change,* MAA Notes, 22, Washington, D.C.: Mathematical Association of America.

[7] Garfield, J. (1992) "Helping Students Learn" in "Teaching Statistics," in L.A. Steen (ed.), *Heeding the Call for Change,* MAA Notes, 22, Washington, D.C.: Mathematical Association of America.

[8] Iversen, G. (1991) "Writing Papers in a Statistics Course," *Proceedings of the Section on Statistical Education,* American Statistical Association.

[9] Lock, R.H. (1990), *Alternative Introductions to Applied Statistics for Mathematics Students,* SLAW Technical Report, 90-008, Pomona College, Pomona, CA.

[10] Moore, D.S. (1991), *Teaching Statistics as a Respectable Subject,* SLAW Technical Report, 91-002, Pomona College, Pomona, CA.

[11] Moore, D.S. (1995), "Current Trends in Statistics Education" (Invited Talk) at Conference for Isolated Statisticians, Chaska, MN.

[12] Moore, D.S. (1996), "Teaching Statistics" (Keynote Presentation) at Trends in Introductory Applied Statistics Courses: Topics, Techniques, Technology, Framingham State, Framingham, MA.

[13] Moore, D.S., Cobb, G., Garfield, J. and Meeker, W.Q. (1995), "Statistics Education Fin de Siecle," *The American Statistician,*. 49 (3).

[14] National Research Council (1989), *Everybody Counts: A Report to the Nation on the Future of Mathematics Education,* Washington D.C.: National Academy Press.

[15] Parr, W.C. and Smith, M. A. (1998), "Developing Case-Based Business Statistics Courses," *The American Statistician,* 52 (4).

[16] Radke Sharpe, N. (1991), "Writing Papers in a Statistics Course," *The American Statistician,* 45(4).

[17] Singer, J.D. and Willett, J.B. (1990), "Improving the Teaching of Applied Statistics: Putting the data Back into Data Analysis," *The American Statistician,* 44 (3).

[18] Stern, A.L. (1995), "A Study in Diplomacy," *Across the Board,* 32 (2).

TABLE OF CONTENTS

Topical Grid for A Casebook for Business Statistics:Laboratories in Decision Making by N. R. Sharpe, Abdul Ali, and Mark E. Potter

Statistics Topics	Case Title
Graphical and Numerical Description of Data	1. *Decisions in New Product Development (A)*
Analyzing Bivariate Data	2. *Decisions in New Product Development (B)*
Confidence Intervals	3. *Risk and Return in World Markets*
Hypothesis Testing: Two-Sample Tests	4. *Decisions in New Product Development (C)*
Hypothesis Testing: Comparison of Proportions	5. *Decisions in New Product Development (D)*
Simple Regression	6. *Relationship between Market Returns and Interest Rates* 7. *Break-Even Time for New Products* 8. *Mutual Fund Flows (A)*
Multiple Regression	9. *Baseball Stadium Age and Attendance* 10. *Mutual Fund Flows (B)*
Time Series	11. *Mutual Fund Flows (C)* 12. *Sales in a Seasonal Industry* 13. *Boston Sunday Globe (A)* 14. *Boston Sunday Globe (B)*
Advanced Topics	15. *Boston Sunday Globe (C)* 16. *Motor Vehicle Fatalities* 17. *Quarterly Sales at Home Depot*

DECISIONS IN NEW PRODUCT DEVELOPMENT (A)

Statistics Topics: **Data Collection**
Data Types
Graphical Displays

Data File: **npd.xls**

Entrepreneurs often use incubator programs for several months to jump-start their new businesses. According to the National Business Incubation Association (NBIA), business incubation is a dynamic process of business enterprise development, where incubators provide hands-on management assistance, access to financing, and orchestrated exposure to critical business or technical support services. Primary sponsors of incubators are nonprofit organizations, such as academic institutions or government agencies. A 1998 study of the Business Incubation Industry done by NBIA reports that North American incubators have created nearly 19,000 companies still in business, and more than 245,000 jobs.

An important decision for start-up companies is whether or not to participate in incubator programs to assist in the development of a product. Alternatives for these start-ups is to pursue "angels" – investors who dig into their own pockets to give entrepreneurs their first financial lift or venture capitalists. However, venture capitalists are usually agents for other people's money and want to see evidence of viability before investing in companies.

Developing a new product is challenging. Today, product development managers face intense pressure to bring world-class products to market in record times. Many factors contribute to this pressure, including acceleration in the rate of technological development; improved mass communication; more intense competition due to maturing of markets and globalization; and a fragmentation of the marketplace due to changing demographics, shorter product life cycles, and the escalating cost of research and development.

Getting new products to market quickly is critical for a company to gain competitive advantage in the battle for market share. In fact, a shorter time to develop a new product can 1)increase sales through extended product life, 2) increase market share through pioneering, 3) increase profitability through pricing freedom and economy of scale, and 4) enhance a company's image as an innovation leader (e.g., Amazon.com, Apple, Honda, 3M and Wal-Mart). An article published in Fortune magazine (Feb. 1989) noted the following:

> *An economic model developed by McKinsey and Co. suggests that in a market with 20% annual growth rate and 12% price-drop per year, high-tech products that ship to market six months late, but on budget will earn 33% less profit over five years. In contrast, coming out on time and 50% over budget cuts profits only 4% in the same market. (p. 54)*

This case was prepared by Professor Abdul Ali and Professor N. R. Sharpe as a basis for class instruction and discussion. The authors acknowledge the help of Prof. Robert Krapfel and Prof. Doug LaBahn in collecting the survey data.

To improve profitability and gain competitive advantage through new product development, the message is consistent from researchers and practitioners alike – accelerate development speed, reduce product cost, improve product performance, and cut development program expenses. All of these recommendations seem reasonable and companies like Hewlett-Packard Co., Honeywell Inc., Intel Corp., and Xerox Corp. have reported significant reductions (as high as 50%) in product development time, accompanied by lower development cost, improved product quality and increased market share. However, such dramatic results are yet to be realized in many small companies. If a company has limited resources (like most start-ups), one needs to make trade-offs between these four possible objectives. How does one go about choosing which trade-off to make?

To help with such decisions, a survey was administered to collect data about new product development practices from small firms, some of which had worked within business incubator programs. The main objectives of this survey were:

- to identify organizational characteristics of small firms that have participated in incubator programs;
- to compare length of development time, price, and competitive characteristics of the product across firms;
- to assess first-year market performance of new products for firm participants and non-participants in incubator programs.

Existing questions are: Is it an advantage for most companies to speed up the new product development process? What factors influence development time for developing new products and what are the influences of development time on market performance? Does it make a difference whether a company participates in an incubator program or not?

Data Collection

The **sampling frame** selected for the study consisted of a wide cross-section (9 different 4-digit Standard Industrial Classification[1], or SIC, groups) of small (less than 100 employees) manufacturing firms. The sampling frame was constructed from three sources. First, a random sample was drawn from a highly regarded commercial mailing list provider. Second, a complete **census** of the Small Business Innovation Research (SBIR) phase II award winners was obtained. Third, a list was compiled from names submitted by a census of the members of the leading association of small business incubator directors.

[1]Standard Industrial Classification (SIC) coding scheme was developed by the federal government to make it easier to collect and tabulate statistics on products, industries, or services (e.g., automotive, computer and data processing services, electronic and electrical equipment, health services, heavy construction equipment), especially in the various economic censuses. Now, many commercial sources present industry data using the different SIC groups.

The survey sponsor chose these firms to represent a wide variety of small businesses engaged in product development to investigate broad patterns of new product development activities independent of industry specific practices. The **unit of analysis** was the firm's most recently completed new product development project. Entrepreneurs (e.g., president or owner) were used as single key informants, since it was presumed that they had vested interest and intimate knowledge of their firms' new product development processes. The pretest interviews and discussions with industry experts confirmed the owner's knowledge and accessibility of the development process.

The data collection phase proceeded in four stages. First, the survey administrators undertook unstructured personal interviews with several entrepreneurs and industry experts. The interviews focused on identifying the most important issues facing the key decision-maker. During this stage, it was noted that new product development is an infrequent activity in many small firms; at any one time many firms are unlikely to have recently completed a project. This was evidenced by several hundred responses indicating that the firm had not recently developed a new product.

The second stage of the data collection process was the development of the survey, which was based on a literature review and feedback obtained during the personal interviews. The qualitative feedback focused on the content and wording of the measurement indicators to minimize measurement error. The third stage was the mailing of 3071 invitations to potential participants in the study, of which, 592 (19.3%) executives agreed to participate. Since the unit of analysis was the firm's most recently completed new product development project, only those firms that had recently completed, or were close to the completion of the product development project were considered for the research. There was no regional bias among those who agreed to participate in the survey. This approach enabled the survey researcher to construct a list of firms that had recently developed a new product. In addition, this approach enabled them to identify the key decision-makers and to gain their commitment.

The final stage of data collection was the actual mailing of the surveys (see **Exhibit 1**) to those who had agreed to participate in the study. Once these surveys were returned, the information was coded and entered into Excel to be used for **descriptive** and **inferential** analysis.

Non-response bias was assessed by comparing early versus late respondents as suggested by Armstrong and Overton (1977). The time between when the questionnaire was mailed and when it was returned was used to form early (67%) and late (33%) respondent groups. Subsequent statistical tests revealed that no significant differences existed between the groups regarding company age, staff size, new products recently launched, and number of products currently sold. Therefore, non-response did not appear to be a major concern in this survey. Sixty non-respondents were also contacted by telephone in order to determine the reason for nonparticipation. The majority reported that they had not recently developed new products.

<u>**Data Analysis**</u>

Of the 592 questionnaires mailed, 286 (48%) were returned and 246 (42.0%) were largely completed (see **bar chart** in **Figure 1**, created in Excel). We will consider this latter group of survey respondents as our usable **sample**. One hundred and twenty-nine of these firms (53%) provided information about a recently completed project, whereas 114 (47%) provided information about an advanced on going product development project. Three companies didn't provide any information regarding their project completion status.

Figure 1

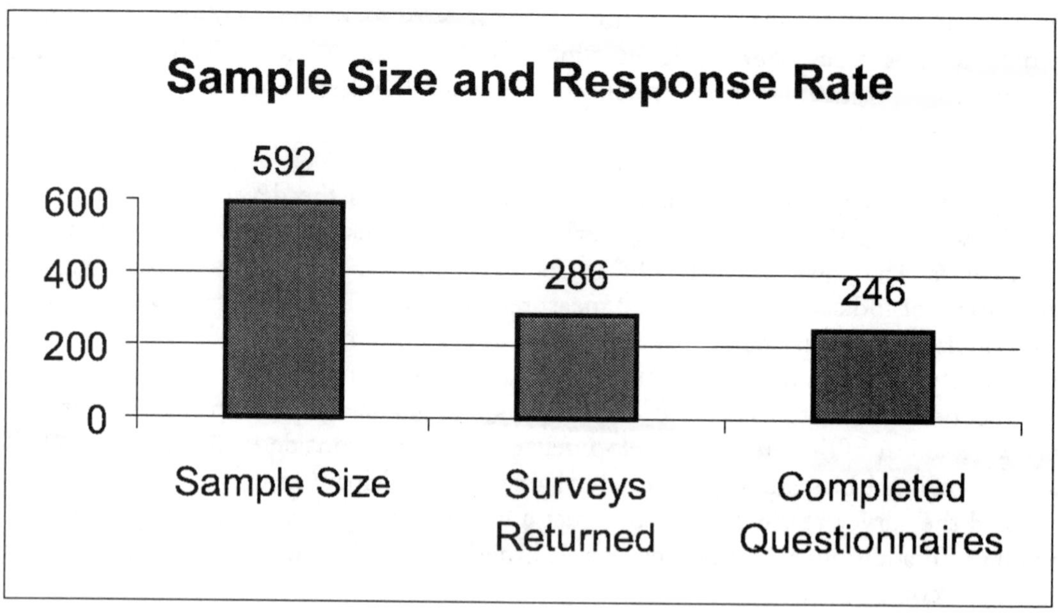

To summarize the data and get a better understanding of the characteristics of the survey, we need to analyze the data. The appropriate analysis technique depends on the **type of data**, and any **inference** is constrained by the type of information collected in the survey, as well as the way the survey was administered. There are four main sections to the survey: section A obtains industry information; section B asks respondents to provide project and launch strategy they undertook to develop new products; section C seeks information concerning the outcome of the development projects; and section D is designed to understand the organization of the respondents.

To observe graphically the relative percentage of incubator participants, we can create a **pie chart** (see **Figure 2**, created in Excel). Of the 246 respondents, 29% participated in an incubator program of some kind. Among those who participated in an incubator program, 35 firms participated in non-profit incubator programs.

Figure 2

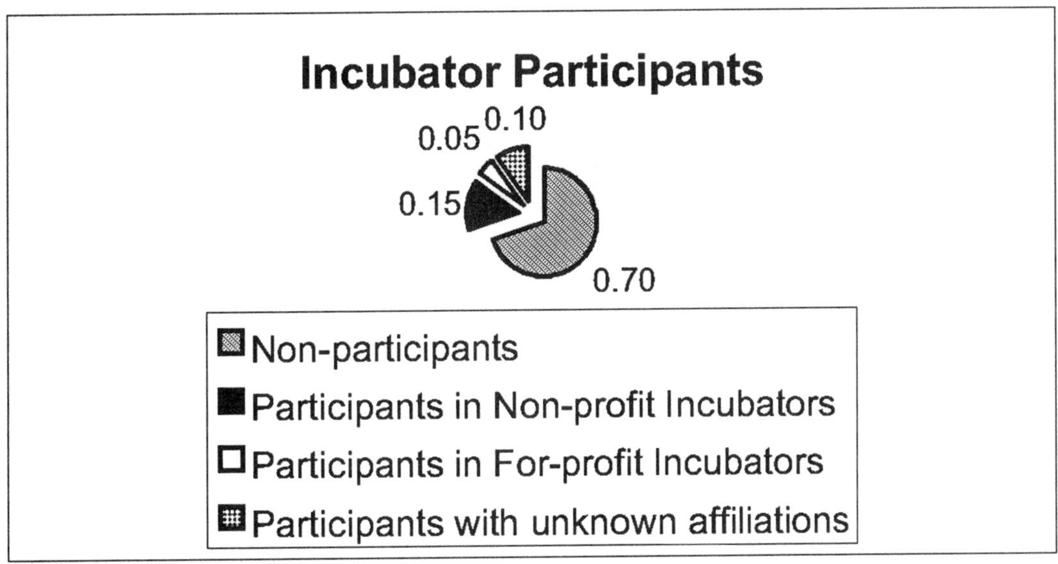

The variable of total annual sales (question 8 in section D) is an indicator of company size, and is an important variable that affects new product development. A **bar chart** of company sales level is shown in **Figure 3**. As you can see, the majority of company sales are less than $1 million.

Figure 3

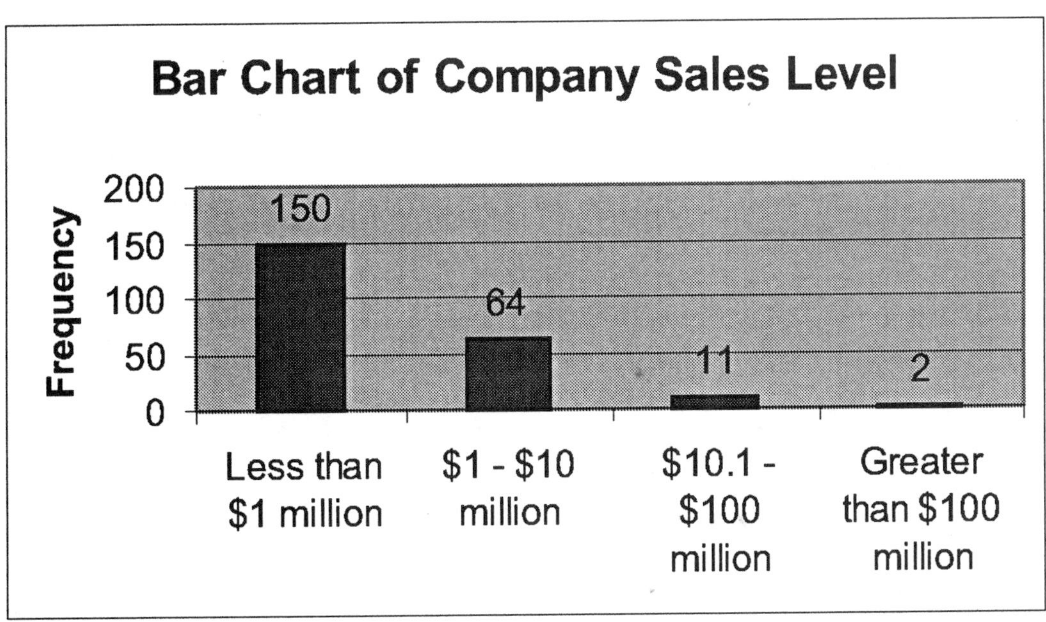

We have used charts to describe only a few of the variables created from the survey. In the questions below, we ask you to use your knowledge of data, charts, numerical measures, and distributions to develop a complete description of the organizational characteristics of the sample.

Conceptual Questions:

1. Examine the measurement scales used in the survey. What are the strengths and weaknesses of the different scales?

2. Which of the survey responses are **qualitative**? What types of charts are appropriate for describing these variables?

3. Which of the survey responses are **quantitative**? **Continuous**? What types of charts are appropriate for describing these variables?

Analysis Questions:

1. For questions 2 and 3 in Section D (variables D2 and D3) create charts to investigate the responses. Describe the sample in terms of their responses to these two questions. (i.e., What percentage of firms are privately held? What percentage of firms received venture capital as a source of funding?)

2. For questions 1 and 4 – 7 in section D, create an appropriate graph to investigate frequency distribution. Which variables have symmetric frequency distributions? Skewed distributions? For those variables with skewed distributions, write a sentence interpreting the skewness in the context of the variable and survey.

3. For the same variables you used in question 2 above, generate summary numerical measures (i.e., mean, mode, median, trimmed mean, standard deviation, and variance).

4. Based on your responses to questions 1 through 3, above, summarize how the majority of respondents answered the demographic (Organizational Characteristics) questions? (i.e., What is the average age of the firms in the sample? What are the most common sources of funding? What is the average size of the firms in the sample? What is the distribution of total R&D professionals? What is the range of the number of product offerings?)

Sources:

Armstrong, J. Scott and Terry S. Overton (1977), "Estimating Non-response Bias in Mail Surveys," *Journal of Marketing Research*, 14 (August), 396-402.

Dumaine, B. (1989), "How Managers can Succeed through Speed," *Fortune* (February 13), 54-59.

http://www.nbia.org/

Exhibit 1: NEW PRODUCT DEVELOPMENT PROCESS

Section A: Industry and Competitive Conditions

The following questions request background information useful in understanding the competitive environment you face in bringing out a new product. In your responses please focus on your most **(i) recently completed** (in the last 2 years) or **(ii) advanced on-going** product development project.

1. Please identify the general product category this new product falls in:______________________________

2. Has the project been completed? ______Yes ______ No (please check one)

3. Approximately how many customers are there for your product? ______ customers

4. Approximately how many competitors are there in this product category that you are competing with? ________competitors

For the following questions, please circle the response number which best describes your situation.

5.	How easy is it for customers to substitute other products in place of yours?	Easy	1 2 3 4 5 6 7	Difficult
6.	Many competitors have introduced competing products in the last 2 years.	Disagree	1 2 3 4 5 6 7	Agree
7.	The short-term (3-yr.) market growth rate for your product is	Low	1 2 3 4 5 6 7	High
8.	The average industry pretax profit level is	Low	1 2 3 4 5 6 7	High
9.	How long will the current technology be viable in the market?	Few yrs	1 2 3 4 5 6 7	Many yrs
10.	What is the average length of time between product/model changes?	Few mths	1 2 3 4 5 6 7	Many mths
11.	What is the prospect for future profits from your product?	Low profits	1 2 3 4 5 6 7	High profits

---

Section B: Product/Launch Strategy

The following questions request information useful in understanding the strategy you undertook (will undertake) in launching the new product into the market. Please continue to focus on the **SAME PROJECT** as in sections A and B. If your responses are about a recently completed project, the answers in this section should solely be based on what you had and did in the initial stage of the market launch of your product. If your project has not been completed please provide your best estimates.

1. Was your company the first into the market with this type of product? ____ yes, ____ no

2. The estimated unit price of the product $_______________

For the following questions, please circle the response number which best describes your situation.

3.	Compared to competitors, the new product meets (will meet) customer needs better.	Disagree	1 2 3 4 5 6 7	Agree
4.	Compared to competitors, the new product was (will be) priced lower.	Disagree	1 2 3 4 5 6 7	Agree
5.	Compared to competitors, the promotional budget of the new product was (will be)	Lower	1 2 3 4 5 6 7	Higher

6.	Compared to competitors, the number of distribution outlets was (will be)	Lower 1 2 3 4 5 6 7 Higher
7.	The new product was (will be) priced much higher than industry average	Disagree 1 2 3 4 5 6 7 Agree
8.	The promotional effort was (will be) much above the industry average.	Disagree 1 2 3 4 5 6 7 Agree
9.	The number of distribution outlets were (will be) much above the industry average.	Disagree 1 2 3 4 5 6 7 Agree
10.	The new product was (will be) of higher quality than any existing product.	Disagree 1 2 3 4 5 6 7 Agree
11.	The new product had (will have) unique features/attributes from existing products.	Disagree 1 2 3 4 5 6 7 Agree
12.	The price of the new product was (will be) higher than existing products.	Disagree 1 2 3 4 5 6 7 Agree
13.	The promotional expenditures did (will) exceed those on existing products.	Disagree 1 2 3 4 5 6 7 Agree

---

Section C: Project Outcomes

The following questions request information concerning the outcomes of your product development process. In responding to this section please concentrate ONLY on the **SAME PROJECT** described above. The beginning of idea generation should be when the firm decided to develop a new product. The end of market launch refers to when the product is commercially available and managed in a routine manner. (For those who have not yet launched a new product please provide your best estimates.

1. Total project time from the beginning of idea generation to the end of the market launch? ____________ months

2. Total project cost from the beginning of idea generation to the end of the market launch? ____________ dollars

3. Total man-years used from the beginning of idea generation to the end of the market launch? ____________ man-years

4. Total time from market launch to attaining break-even? _________ months

5. We are satisfied with our product development process. Disagree 1 2 3 4 5 6 7 Agree

6. We would use the same product development process over again. Disagree 1 2 3 4 5 6 7 Agree

7. We would recommend the same product development process to others. Disagree 1 2 3 4 5 6 7 Agree

8. What was the actual (or expected) first year sales level? (please check one)

____ less than $100,000	____ $ 1 to 5 million
____ $100,000 to $499,999	____ $ 5 to 10 million
____ $500,000 to $ 1 million	____ more than $ 10 million

9. What was the actual (or expected) first net income as a percent of sales? (please check one)

____ loss	____ 11 to 20%
____ 0 to 5%	____ 20 to 30%
____ 6 to 10%	____ over 30%

Section D: Organizational Characteristics

The questions in this section are designed to help us better understand and classify the information provided above. This information will be treated confidentially and only aggregate statistics will be reported. **No information from any one respondent will be released to anyone.**

1. What year was the company founded? Year 19____

2. Is this company privately held or publicly traded? (please check one)

   ____ privately held
   ____ publicly traded

3. Which of the following sources of funding has this company received? (please check all that apply)

   ____ Venture capital
   ____ Bank loan
   ____ SBIR
   ____ Initial public offering
   ____ Private placement
   ____ Other public sector grants
   ____ Other (please describe) ____________________

4. What is the size of your organization in terms of total full time staff? _______ Total staff

5. Of your total staff, how many are R&D professionals? _______ R&D professionals

6. Number of other new products launched in the last **two years** _______ new products introduced

7. Total number of product offerings now sold _______ products currently sold

8. Which of the following categories best represents total yearly company sales prior to launching the product? (check one)

   ____ Less than $1 million
   ____ $1 to $10 million
   ____ $10 to $100 million
   ____ Greater than $100 million

9. As a start-up company, did this organization participate in an incubator program of any type?

   ____ Yes --> If yes, was the incubator ____ non-profit ____ university
   ____ for-profit ____ government

   ____ No

We greatly appreciate the time and effort you have spent on our survey. Please take a moment to complete any questions you may have accidentally skipped. If you would like to receive an executive summary of the survey results, please attach your business card to the questionnaire. Finally, please return the questionnaire in the enclosed return envelope.

Thank You

DECISIONS IN NEW PRODUCT DEVELOPMENT (B)

Statistics Topics:	**Bivariate Data** **Contingency Tables** **Scatter Plots**
Data File:	**npd.xls**

Many small firms must decide whether or not to use the assistance of an incubator program to develop a product. This is the second case in a series that discusses issues in the new product development process and investigates the relationship between incubator programs and product development outcomes. In Part A of this case, we described the data collection process used to obtain information about new product development practices of small firms. The survey is displayed in Exhibit 1 of Part A.

In this case we continue the analysis of the product development survey data. Our main objective is to investigate the relationship between project outcomes (i.e., first-year sales level, net income as percent of sales, project cost, project development time, and length of time from market launch to breaking even) for the two groups of respondents: incubator participants and non-participants.

Exploratory Analysis

Since participation in an incubator program can be competitive, it is natural to hypothesize that participation in such a program should provide a start-up with some advantages. Thus we are interested in examining if more participants report higher first-year sales than non-participants. From the pie chart in Figure 2 of Part A of this case, we know the frequency distribution of incubator participants and non-participants. **Table 1** is a **contingency table** for two variables: participation (D9) and first-year sales from the product developed (C8). Note, that the contingency table below shows the number of respondents as 214, which is less than the total number of survey respondents (246). This is because the respondents who did not respond to either question used to construct the contingency table (i.e., incubation participation or first-year sales) produced missing observations, and thus could not be included in the table.

From Table 1, it can be seen that no incubator participant reported more than $10 million in sales in the first year of launching its new product. We can also observe that the majority of *both* participants and non-participants reported sales of under $500,000. However, further meaningful comparisons between the two groups are difficult from examining the frequencies. For further conclusions, we ask you to examine the **relative frequencies,** using either the total percentage, the row percentage, or the column percentage in the analysis of this case.

This case was prepared by Professor Abdul Ali and Professor N. R. Sharpe as a basis for class instruction and discussion. The authors acknowledge the help of Prof. Robert Krapfel and Prof. Doug LaBahn in collecting the survey data.

Table 1

Contingency Table for First-Year Sales Level and Incubator Participants							
	Less than $100,000	$100,000-$499,999	$500,000-$1 million	$1.1million - $5 million	$5.1million –$10million	More than $10million	**Row Total**
Incubator Participants	24	22	8	6	1	-	61
Non-Participants	53	61	13	19	2	5	153
Column Total	77	83	21	25	3	5	214

For a company, in addition to the first-year sales level of its new product, it is equally important to know the percent of sales which is profit for the company. Both first-year sales level (revenue) and income as percent of sales (return on sales) are important measures of the market performance of a new product. Hence, we next use the **clustered bar chart** to analyze if participation in an incubator program makes a difference in making profit from new product development (first year net income as percent of sales from new products).

The clustered bar chart in **Figure 1** (created in Excel) below shows the result. From this chart, we can see that except in one range of profit, incubator participants are not reporting better performance than non-participants. In fact, 32% of incubator participants report a loss, while only 16% of non-incubator participants report the same. Similarly, only 17% of incubator participants report their new products are contributing the highest level of revenue as income, while 22% of non-incubator participants report the same. What could be the reason for this surprising result? One possibility is that incubator participants report higher losses because they accelerated the product development process by incurring a higher product development cost. If this is true, then we would expect there to be an **inverse relationship** between development time and development cost. How should we investigate this potential link between increased project cost and shorter product development time?

First, we examine if development time is related to project cost by using a **scatter plot**. **Figure 2** (created in Minitab) shows a scatter plot of project cost (C2) versus project development time (C1), where the two groups – participants (P) and non-participants (NP) – are identified by different symbols. What can we learn from Figure 2? Note that there are several observations unusually high in project cost (above $20 million). Are these unusual observations all members of the same group (participants or non-participants)? How can this result affect the results observed in Figure 1? Why?

Figure 1

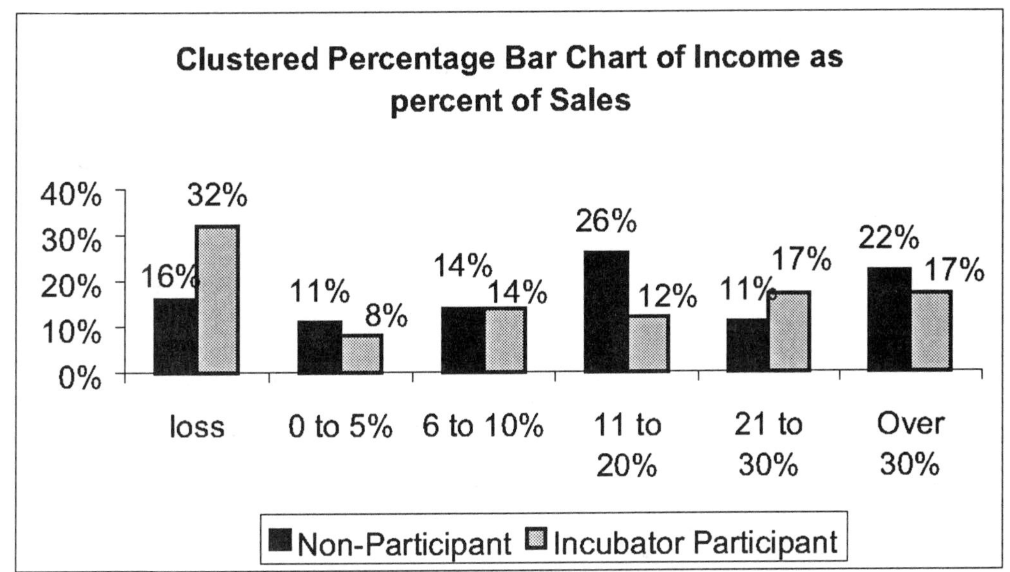

Figure 2

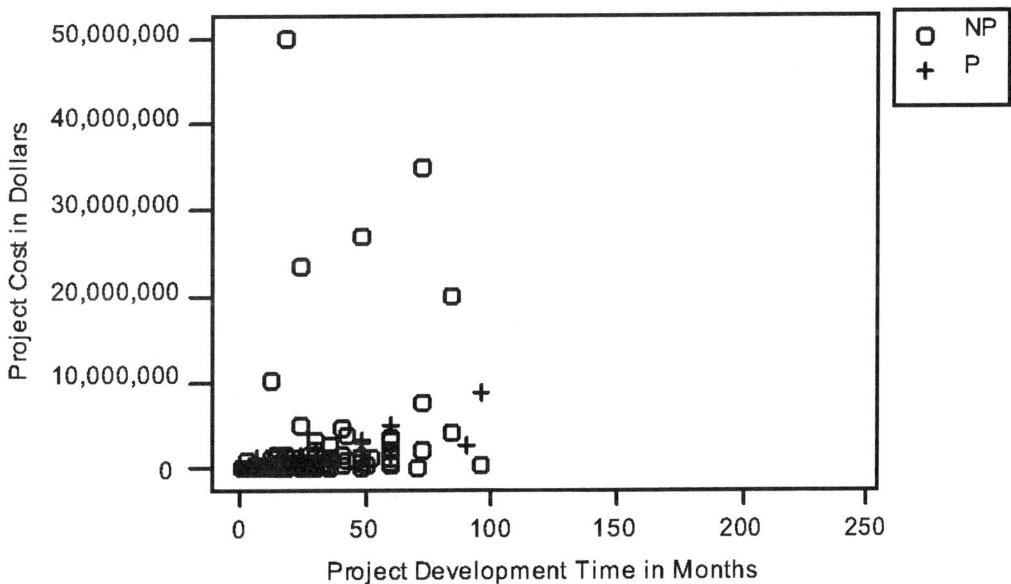

These unusual observations may be obscuring the true relationship between project cost and project time, because they are not typical of the sample. Thus we decide to remove these unusual observations from the analysis. **Figure 3** is a scatter plot of project cost versus project development time omitting those five companies that spent more than $20 million.

Figure 3

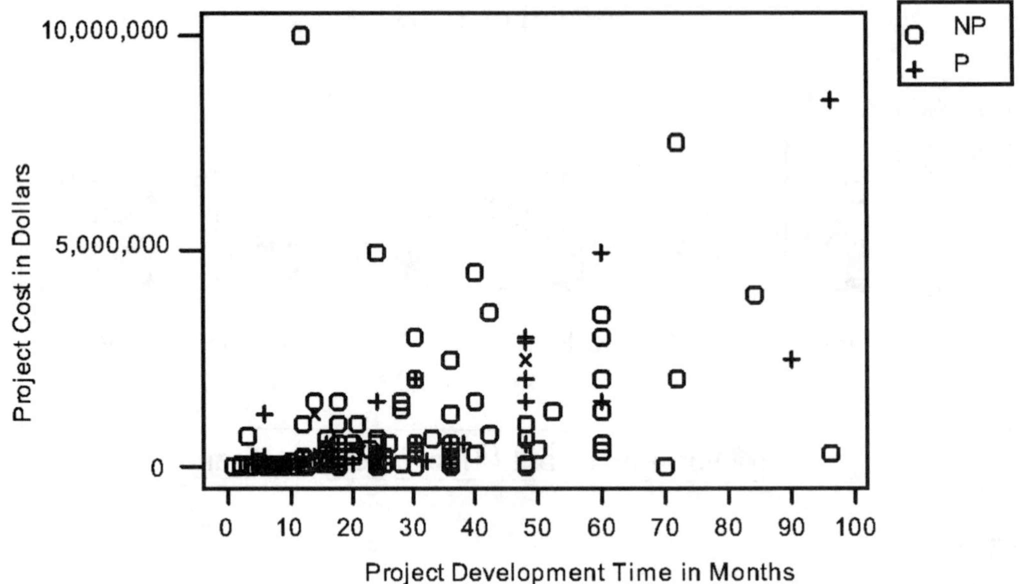

Does this graph show more clearly the relationship between project cost and development time? Do you suspect that this relationship is the same for both non-participants and participants? Do you continue to see more variability in the graph than you would suspect for a "strong" **linear relationship** to exist between these two variables?

Transformation of Variables

There is also a possibility that the relationship between these two variables is non-linear. One indication of a non-linear relationship is a skewed distribution of the variables. (We ask you to verify the distributions of these continuous variables using graphical displays in the analysis of this case.) One technique often used to normalize skewed data is a logarithmic transformation of the variable. Thus we use a logarithmic transformation of both project cost and development time. **Figures 4 and 5** (created in Minitab) show the resulting scatter plots between the two transformed variables (omitting companies spending greater than $20 million) for the two groups – non-participants and participants.

Compare Figures 4 and 5. How are they different? How are they similar? These graphs show that total project cost may have a positive relationship with project completion time – for both participants and non-participants; if a new product development project takes longer to complete, it is expected to cost more. Therefore, it does not appear that a higher cost, resulting in shorter completion time, explains the poor performance observed in Table 1 and Figure 1 for many companies that participated in incubator programs.

Figure 4

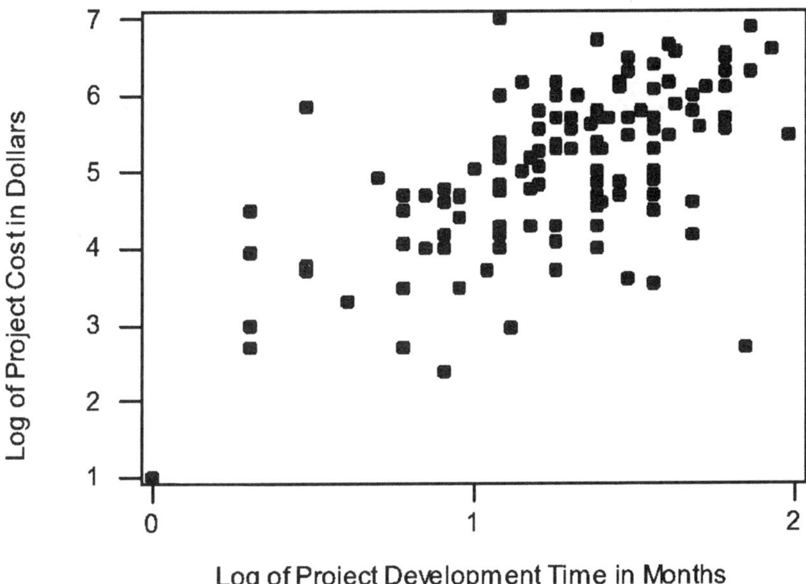

Figure 5

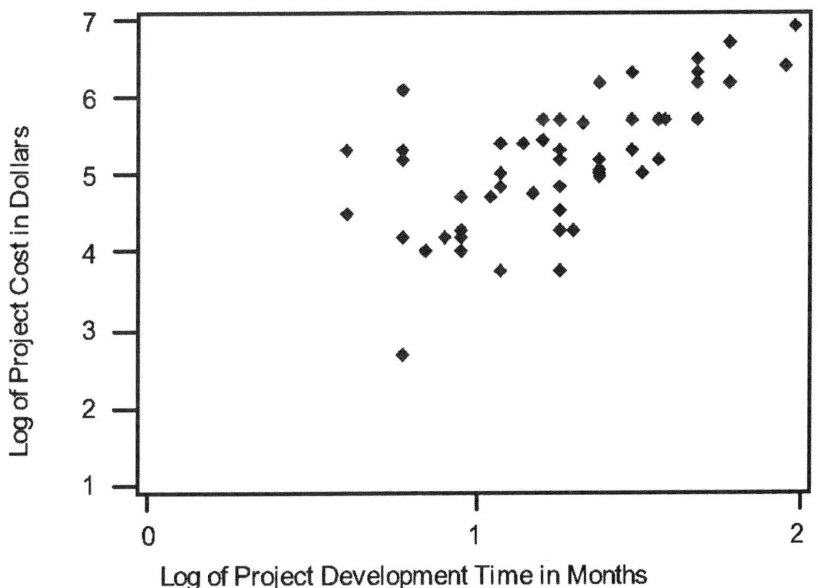

Is there an alternate possibility? Could incubator participants report higher losses because they took a longer time to break even with their new products than non-participants? Participation in an incubator program may help a firm, who has lack of experience and resources in developing a new product, but may not provide much help in launching the new product. What do you think? We ask you to investigate this theory below in the case analysis.

Analysis Questions

1. For the data in Table 1, compute the row, column, and total percentage for each cell. For a contingency table, why do you need to calculate row percentage or column percentage? How does this help with interpretation?

2. Create a clustered bar chart for income as a percentage of sales *not* including those firms reporting a cost of greater than $20 million. How is this chart different from Figure 1? Does this alter the comparisons between participants and non-participants?

3. Investigate the distribution of project outcome variables development time (C1), development cost (C2), and break-even time (C4). Which ones have a symmetric

distribution? Which ones have a skewed distribution? For those that have a skewed distribution, transform using logarithms and create a histogram for the transformed variables. Is the distribution more symmetric?

4. Create a scatter plot for the project outcome variables break-even time (C4) and project development time (C1). Identify whether the observations are members of the participant or non-participant group. Are there any firms that appear unusual in either break-even time or development time? Do you observe a relationship between these two outcome variables in the scatter plot? Do you suspect that this relationship is the same for the two groups? Do you suspect that incubator participants have a longer break-even time than non-participants?

Discussion Questions

1. As shown in Figure 1, why do you think a larger proportion of incubator participants reported a loss than non-participants? Do you think it suggests that incubator programs have a negative effect on new product development? Or, can you think of another possible explanation for this relationship? Should this relationship affect our recommendation to managers? Would it discourage us to recommend to mangers that their company should not participate in an incubator program?

RISK AND RETURN IN WORLD MARKETS

Statistics Topics:	**Histograms** **Numerical Summary Statistics** **Measures of Central Tendency and Variability** **Confidence Intervals**
Data File:	**markets.xls**

The world of investments is attractive to an increasing proportion of today's society. And why not – the prospect of turning a small amount of savings into a large pile of cash is appealing. The concept seems simple. Rather than put your savings into the bank, put it into the stock market. We are constantly barraged with commercials for brokerage firms, mutual fund companies, and day-trading firms, all of which entice the eager consumer with promises of early retirement, a house on the ocean, and cruises around the world. Few of us hear the disclaimers that usually appear at the end of the advertisements. They consist of statements like, "past performance may not be indicative of future results," "performance is not guaranteed," and of course, "there is a chance of substantial loss of wealth." The nature of greed comes into play for many investment decisions. That is, the riskier the investment, the higher the returns. Those unwilling or unable to take on risk are destined to receive low returns on their investment. Does this have to be the case, or are there investments where you could earn a handsome return with minimal risk?

Examining Return

There are many vehicles that can be used for investing, and some are riskier than others. For example, the United States government guarantees money in savings accounts. This means that if you put your cash into a bank savings account you cannot end up with less money than you started with; that is, unless you withdraw it to purchase concert tickets, a sports car, or a Mark McGuire rookie card. Because savings accounts are so safe, they do not pay much interest. Currently, savings accounts pay about 3 percent per year. So, if you put $1,000 into a savings account at the beginning of the year and keep it there for a whole year, the bank would pay you $30, giving you $1,030 at the end of the year.

Many people are not satisfied with 3 percent and want to invest in the stock market, where returns can be higher. Of course, this also means that the risk is greater, which means that returns can also be lower, or even negative. But what does "invest in the stock market" really mean? There are different methods of investing in the stock market. You could open an account with a broker for several thousand dollars, and instruct the broker to purchase shares of stock in your favorite companies, like Microsoft, Intel, Disney, or Amazon.com.

Shares of stock represent ownership, so if you own a share of stock in a company, you own a piece of the company (albeit a small piece). If the company sells a lot of software, has a winning season, or sells a lot of books, then the company will be worth more over time, and the shares of stock you own will increase in value. If the company loses a legal battle, cannot control its costs, or loses sales to a new competitor, then the company's value (and your shares of stock) will decrease in value over time. Typically, investing through brokers is a medium- to long-term method of investing in the stock market.

Another way of investing in the stock market is through "day-trading." This is essentially a short-term investment technique. The investor sits at a computer terminal and typically buys and sells many times in a single day, hoping to make money on emotional swings of the stock market. Because most stock prices don't move much in a single day, it takes a lot of money for day-traders to earn a living after paying trading expenses. It is also, as you can imagine, a risky way to invest in the stock market, and many reports have come out comparing day-trading with gambling.

A third way of investing involves mutual funds. These are portfolios of stocks that are professionally managed by a company, such as Fidelity or Vanguard. An investor would pay for a share of the mutual fund, which represents a piece of the portfolio. Many times, there are hundreds of stocks owned by the mutual fund, so even if you owned only a few shares of the mutual fund, you would own a piece of those hundreds of stocks. This reduces your risk through a concept called "diversification", meaning that even if some of the stocks owned by the fund decline in value, others will probably go up and they will even out over time.

Suppose you decide that you want to invest in the overall stock market. This can be done, but which stock market do you want to invest in? There are many all over the world, from the United States to Russia to Hong Kong to Australia. The performance of these markets varies widely. For example, in 1995, the Swiss stock market grew about 35%, while the stock market in Japan grew by less than 3%.

Table 1 on the next page lists the major world stock markets and their respective returns during the calendar year 1995 (in alphabetical order). Note that some are much higher than others. If someone asked you which market had the best return, you could scan the entire list and find the market with the highest return, which would be Switzerland. If someone asked you how many stock markets earned more than 10% during 1995, you would again scan the list, counting the number as you go along. A much more efficient and effective mechanism for viewing and exploring data is to display the data graphically.

Table 1 Stock Market Return in 1995 for different World Markets

Country	Market Return (%)
Australia	15.92
Austria	-0.69
Belgium	22.51
Canada	16.80
Denmark	16.76
France	10.16
Germany	15.95
Great Britain	20.04
Hong Kong	24.54
Italy	-0.84
Japan	2.85
Malaysia	6.31
Netherlands	26.42
Norway	9.46
Singapore	8.78
Spain	24.75
Sweden	33.34
Switzerland	34.54
United States	32.69

Describing and summarizing data numerically is one of the most important application of statistics in business. The concepts of central tendency and variability of data are used in all business disciplines, including marketing, accounting, finance, operations, and human resources. To this point, you have probably seen a number of graphical representations of data. For example, the **histogram** below (**Figure 1**) shows the frequency distribution of returns of the United States stock market[2] over the past twenty-two years.

[2] The returns are actually from an index of 500 companies, called the Standard & Poor's 500 (S&P500) that was designed to represent the entire United States stock market.

Figure 1

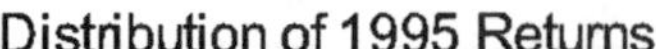

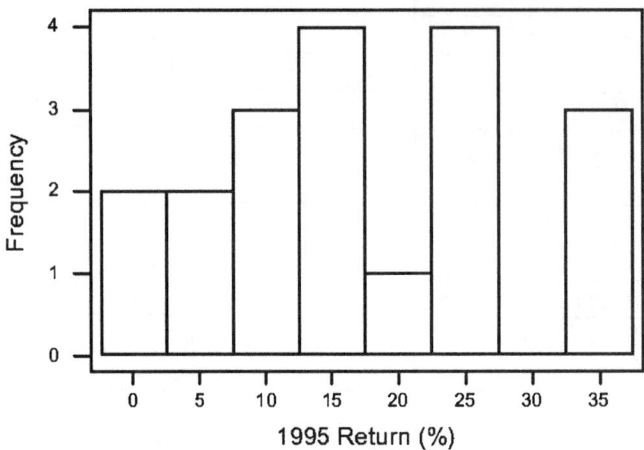

While the figure above is adequate for providing a picture of how the returns are distributed, using some basic summary statistical measures, we can learn even more about the data to help us make a decision. What type of decision? Well, what if your uncle (a sophisticated investor) asked you about how the United States stock market has performed in the past twenty years compared to the stock market in Switzerland, his birthplace? We could graph the two side by side in the form of a **clustered bar chart**, as seen below in **Figure 2**, and draw some comparisons regarding performance across the two markets.

Figure 2

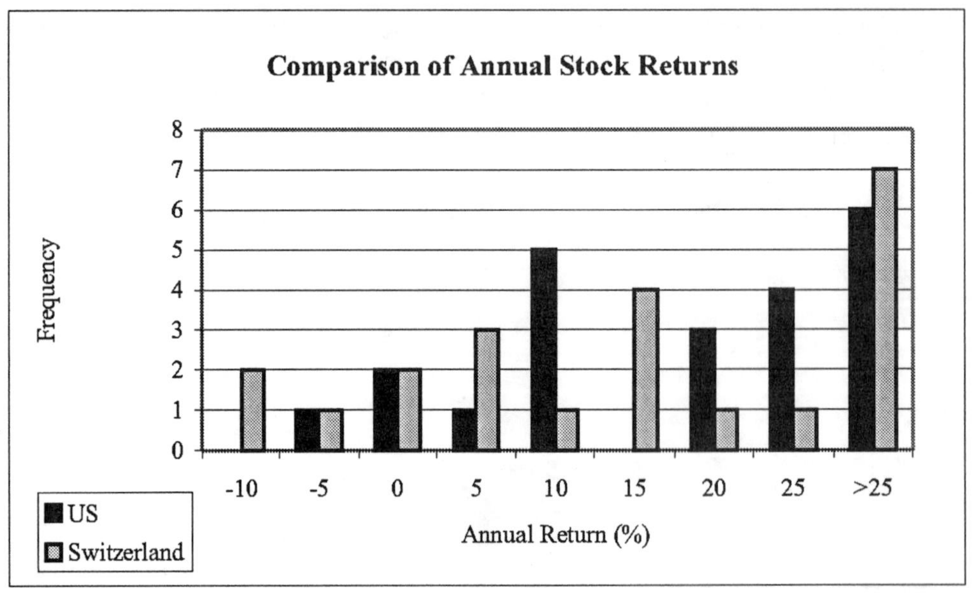

Risk Analysis

As the old saying goes, "there's no such thing as a free lunch." This essentially means that if you see something that is too good to be true, it probably is. You found out from the analysis above that some stock markets have historically earned great returns, and others have not. **Table 2** below shows summary statistics for the stock markets in the United States, Australia, Japan, and Europe. The data represent *monthly* returns from January 1975 through December 1996.

Table 2 Summary Statistics for Stock Market Return (%)

	Europe	Australia	US	Japan
Mean	1.29	1.32	1.30	1.31
Median	1.34	1.15	1.35	0.88
Standard Deviation	4.78	7.03	4.19	6.65
Maximum	23.02	22.23	13.31	25.88
Minimum	-18.87	-44.39	-20.77	-18.50
Range	41.89	66.62	34.07	44.38
Count	264	264	264	264

As you can see, there is considerable variability in the data from one stock market to the next. For example, the median United States market return was 1.35 percent per month, whereas the Japanese median monthly return was just 0.88 percent. This may not seem like a big difference. However, if you had invested just $200 per month from 1975 to 1996 into each market, you would have ended up with almost $290,000 *more* in the United States stock market account than in the Japanese stock market account!

Analysis Questions:

1. Construct a **frequency table** and **relative frequency** table. Using information provided by the tables, draw conclusions about stock market performances overall in 1995. (i.e., Did most world markets have at least 15% return? Was the performance of some of the markets noticeably different from the others? What was the most common range of performances for these markets?

2. Examine the distribution graphically for the 19 world markets in Table 1. Now, that you have a visual view of the distributions of the market returns, does this enhance your response to question 1?

3. Using the monthly data provided in the data set, examine the central tendency and variability for each world stock market in Table 1. (i.e., Compute the summary statistics, such as mean, median, standard deviation, etc.) What do these numbers represent? How would you describe the return performance in each market? Does the answer differ depending upon which measure of central tendency you use?

4. Rank the stock markets from high to low using the mean monthly return. Does it appear that the stock markets that had the highest returns also contained the most risk? Which markets appear "out of place"?

5. Create the frequency distributions for the four markets in Table 2. What information can you obtain from them about risk for each of the markets?

6. An additional means of comparing the risk-adjusted performance of world markets involves the use of confidence intervals, a mechanism for comparing the standard deviation adjusted range for each market. Using both the *annual* and *monthly* returns for the four world markets in Table 2, compute the following confidence intervals:
   a. 90%
   b. 95%
   c. 99%

   Which markets appear to be the most and least attractive using the monthly data? Do your conclusions differ if you're looking at the annual data? What do the confidence intervals tell you about the variability of each market? Did any of the markets yield significantly different returns over the time frame of the data set? At what significance level?

Discussion Questions:

1. What are the benefits and drawbacks of the various types of graphs and charts you created above? Which do you think provides the best way of displaying this particular data set?

2. What are the most striking differences between the markets summarized statistically in Table 2? What are the similarities? Why, for some markets, is the mean similar to the median while for others the mean is very different from the median? What does this imply about the underlying distribution of returns?

3. The standard deviation of stock returns is one way to measure the riskiness of a stock market, as it is an indication of the up and down movement of the market over time. We can observe that the Japanese market has a greater standard deviation than the US market. Does this mean that the Japanese stock market is always more risky than the US stock market? What does this imply about the "no free lunch" argument? Does this argument seem to hold in general for the markets in the table?

4. An additional summary statistic that compares risk and return is called the "coefficient of variation" (CV), which is computed as CV= standard deviation/mean. The interpretation is the units of risk taken on for each unit of return obtained. Compute the coefficient of variation for each world market in Table 2, and compare the risk and return for each of these markets.

Source:

http://www.stanford.edu/~wfsharpe/rets/returns.htm
(Returns in Table 1 and Table 2 were adapted from data provided by Independence International Associates, Inc., found at this website.)

DECISIONS IN NEW PRODUCT DEVELOPMENT (C)

Statistics Topics: **Hypothesis Testing**
One- and Two-Tail Tests
Two-Sample Tests

Data File: **npd.xls**

This is the third case in a series that discusses issues in the new product development process and attempts to relate time spent on development to future sales. Many small firms must decide whether or not to use the assistance of an incubator program to develop a product. To investigate the relationship (if any) between incubator programs and characteristics of new product development, we used a survey to collect data about new product development practices of small firms. The survey is displayed in Exhibit 1 of Part A of this case, which also provided a detailed description of the data collection process.

In Part B of this case, we investigated the bivariate relationship between variables using clustered bar charts and scatter plots. While such graphical methods helped us to gain insights into the relationship between project outcome and demographic variables, we are not sure whether these relationships are statistically significant or not. Will the results be the same if a different sample (of the same size) is drawn from the same population? More importantly, are the inferences we draw from the results valid or will they change from sample to sample?

In this case we continue the analysis of the product survey data. Our main objective here is to investigate if participation in an incubator program makes a difference in project outcome (i.e., product development time, development cost, and break-even time).

Data Analysis

We are interested in seeing if project outcomes for incubator participants are different from those of non-participants. At this point, we are not sure whether an incubator participant would have a better project outcome than a non-participant. Consequently, we sets up a two-tailed hypothesis test with the following null and alternative hypothesis:

H_0: *There is no difference between the mean time it takes to develop a new product for an incubator participant and a non-participant.*

H_A: *There is a difference between the mean time it takes to develop a new product for an incubator participant and a non-participant.*

This case was prepared by Professor Abdul Ali and Professor N. R. Sharpe as a basis for class instruction and discussion. The authors acknowledge the help of Prof. Robert Krapfel and Prof. Doug LaBahn in collecting the survey data.

In this case, we do not know the actual variance or standard deviation of either the participant or non-participant population. However, we assume that our samples of these two populations are random and independent, and that the two populations are normally distributed. If we are not willing to assume that the population variances are equal, then we can use a separate-variance t-test to compare the means of the two populations. Since, we don't know the population standard deviations of the two groups, we use the **sample** standard deviation (**statistic**) as an estimate for the **population** standard deviation (**parameter**). **Table 1** provides the summary statistics for participants and non-participants and **Table 2** provides the output in Minitab for the separate-variance t- test.

Table 1

Summary Statistics of Product Development Time (in months)		
	Group 1: Non- Participants	**Group 2: Incubator Participants**
Mean	26.22	24.81
Standard Error	2.11	2.58
Median	20.50	18.00
Mode	24.00	18.00
Standard Deviation	25.86	19.67
Sample Variance	668.83	387.00
Range	239.00	92.00
Minimum	1.00	4.00
Maximum	240.00	96.00
Sum	3932.50	1439.00
Count	150.00	58.00

Table 2

Two Sample T-Test and Confidence Interval

```
Two sample T for Development Time

INCUBTR        N       Mean      StDev   SE Mean
0 (NP)       150       26.2       25.9       2.1
1 (P)         58       24.8       19.7       2.6

95% CI for mu (0) - mu (1): ( -5.2,  8.0)
T-Test mu (0) = mu (1) (vs not =): T = 0.42  P = 0.67  DF = 135
```

Looking at the output for the two-tailed test in Table 2, we note that the p-value is equal to 0.67, which is greater than 0.05. Therefore, we do not reject our null hypothesis and we conclude that there is no evidence of a significant difference in the mean product development time between incubator participants and non-participants as reported in the survey. We ask you to investigate differences in the means for other project outcome variables in the analysis of this case below.

Conceptual Questions:

1. How can you attempt to verify the assumptions necessary for the separate-variance t-test? Do you believe that these assumptions are valid for the t-test conducted in the case for development time?

Analysis Questions:

1. Is there evidence of a difference in the means for other project outcomes (i.e., development cost and break-even time) for participants and non-participants of incubator programs? Attempt to verify the assumptions. Are there unusual observations that should be removed? (Since break-even time is the only project outcome variable which is measured after the product is launched (unlike development time and project cost), this requires a long-range estimate on the part of the firm. Thus try using only those firms, which have completed (launched) their products for this analysis.) Conduct the appropriate t-test.

2. Is there evidence of a difference in the mean age of a firm (D1), size of a firm (D4), or number of research and development professionals at a firm (D5) for participants and non-participants of incubator programs? Attempt to verify the assumptions. Are there unusual observations that should be removed? Set up the null and alternative hypothesis for the two-tail test. Conduct the appropriate t-test. Interpret the results in terms of the objectives of the case. Can you think of a possible explanation for your results?

3. Since firms are voluntarily (and often competing) for spaces in incubator programs, we hypothesize that those firms using an incubator program are more satisfied with their product development process.
   (a) Set up the null and alternative hypothesis for the **one-tail test** comparing satisfaction with project outcome (section C) across the incubator participant and non-participant groups. (i.e., Compare process satisfaction (question 5), willingness to use same process (question 6), and recommendation of process (question 7) across the incubator participant and non-participant groups.)
   (b) Conduct the appropriate two-sample tests for the hypotheses in part (a). Interpret your results at a significance level of .05.
   (b) Summarize satisfaction with the development process across incubator participants and non-participants.

Discussion Questions:

1. Evaluate the implications of the results for the first two analysis questions above. Do you think that there might exist a relationship between break-even time and the size or experience of a firm?

DECISIONS IN NEW PRODUCT DEVELOPMENT (D)

Statistics Topics: **Categorical Data**
Chi-Square Test
Comparing Proportions

Data File: **npd.xls**

This is the fourth case in a series that examines organizational characteristics and project outcomes, such as product development cost and time, for companies that participated in an incubator program. Many small "start-up" firms must decide whether or not to use the assistance of an incubator program to develop a new product. To investigate the relationship (if any) between incubator programs and characteristics of new product development, we used a survey to collect data about new product development practices of small firms. The survey is displayed in Exhibit 1 of Part A of this case, which also provided a detailed description of the data collection process. Part B of this series used descriptive graphical analysis to compare project outcomes between program participants and non-participants and Part C of this series used inferential statistics to compare differences in continuous outcome variables, such as break-even time, between the two groups.

In this case we continue the analysis of the data collected with the new product development survey. Our main objective is to investigate the relationship between project outcomes considered **categorical variables** (i.e., first-year sales level and net income as percent of sales) for the two groups of respondents: incubator participants and non-participants.

Since participation in an incubator program can be competitive, it is natural to hypothesize that participation in such a program should provide a start-up with some advantages. Thus we are interested in examining if more participants report a higher sales level and net income as percent of sales than non-participants. From the pie chart in Figure 2 of Part A of this case, we know the frequency distribution of incubator participants and non-participants. From the contingency table in Table 1 of Part B using the variables of participation (D9) and first-year sales (C8), we know the number of participants at each sales level.

Recall that this contingency table showed that no incubator participant reported more than $10 million in sales in the first year of launching its product. We also observed that the majority of *both* participants and non-participants reported sales of under $500,000. Because of the scarcity of companies that reported sales over $500,000, we now decide to collapse these six categories into three: sales less than $100,000, sales between $100,000 and $500,000, and sales over $500,000. The contingency table for these new categories appears below (**Table 1**).

This case was prepared by Professor Abdul Ali and Professor N. R. Sharpe as a basis for class instruction and discussion. The authors acknowledge the help of Prof. Robert Krapfel and Prof. Doug LaBahn in collecting the survey data.

Table 1

Contingency Table for Incubator Participation and First-Year Sales Level				
	Less than $100,000	$100,000-$499,999	At least $500,000	**Row Total**
Incubator Participants	24	22	15	61
Non-Participants	53	61	39	153
Column Total	77	83	21	214

For a company, not only does it make sense to look at the first-year sales level of its new product, but it is equally important to know the percent of sales which is profit for the company. Both first-year sales level (revenue) and income as percent of sales (return on sales) are important measures of the market performance of a new product. Hence, we also examine the number of respondents at each level of first year net income from the new product. From a cross-classification table, we note (as we did in Part B from the clustered bar chart) that the majority of respondents are grouped into three categories: a loss, a net income of 0 to 20%, and a net income over 20%. The contingency table for these new categories (Table 2) appears below.

Table 2

Contingency Table for Incubator Participation and First-Year Net Income				
	Negative	0 to 20%	Over 20%	**Row Total**
Incubator Participants	19	20	20	59
Non-Participants	24	74	48	146
Column Total	43	94	68	205

What can we learn from these two tables? How can we determine if the observed frequency at each level for participants and non-participants is what we expect assuming that these two populations are not different with respect to first-year project outcomes? How can we examine if these two populations can be expected to achieve the same first-year outcomes in terms of sales and net income?

Analysis Questions:

1. Create the new variables of first-year sales and net income reflecting the three categories for each variable as described in the case.

2. Compute the observed percentage of participants in incubator programs, who fall into each category of first-year sales. Then compute the observed percentage of non-participants, who fall into each category of first-year sales. Now plot the percentages for each group (participants and non-participants) at each level on the same graph (proportion at each level versus sales level). What do you observe from this graph?

3. Repeat question 1 above for the variable of first-year net income. What do you observe from this graph?

4. Now, use statistics to examine if the proportion of companies at each level of sales and at each level of net income is the same for both incubator participants and non-participants. Set up the hypotheses, find the expected frequency for each cell, and calculate the Chi-square test statistic. At the .05 level of significance, is there a difference in the proportion of companies at the different levels of first-year product performance between those who participated in an incubator program and those who did not?

5. Use statistics to examine if the proportion of companies at each level of total annual sales volume (variable D8) is the same for both incubator participants and non-participants. (Examine the distribution of the data among the levels of sales volume. Note, that because of the paucity of companies who reported sales volume over $10 million, the top three levels should be combined. Thus this contingency table becomes a 2 x 2 table.) Set up the hypotheses, find the expected frequency for each cell, and calculate the Chi-square test statistic. At the .05 level of significance, is there a difference in the proportion of companies at the different levels of annual sales volume between those who participated in an incubator program and those who did not?

Discussion Questions:

1. Interpret your results in question 4 and question 5 above. Is your result to question 4 a surprise? Can your results in question 5 perhaps help explain these results? Can you think of other possible reasons for this difference between participants and non-participants? Would this affect your recommendation that a company should participate in an incubator program?

RELATIONSHIP BETWEEN MARKET RETURNS AND INTEREST RATES

Statistics Topics: **Confidence Intervals**
Hypothesis Testing
Exploratory Data Analysis
Simple Regression

Data File: **rate.xls**

Though there are many stock markets, the vast majority of international stocks are traded on the world exchanges shown below. Some investors, tempted by the prospect of large global returns, trade stocks on these exchanges and find out that is easier said than done. For example, how exactly do you buy a share of stock on the Netherlands Stock Exchange? Do you call up a traditional broker? Should you exchange your currency over to *Gulden* first? Do they allow foreign investors to purchase the stock you want to buy? Are there ownership or trading restrictions?

In addition to navigating the international constraints on trading, investors are typically interested in learning which economic indicators can be used to track market performance. While there are a variety of indicators that can be used explain how the market has performed, the primary purpose of this case is to examine the relationship between national interest rates and the performance of a country's stock market.

Investing in an emerging market is one of the riskiest forms of investing. It is for this reason that most individuals and fund managers invest in only the major world markets. Furthermore, even among the major markets, stock performance varies widely. Historically, some have consistently outperformed others. **Table 1** lists the major world stock markets, along with some summary statistics, presented on an annual basis.

What are the reasons for one stock market to outperform another stock market? Most investors believe that a stock market performs well when the economy is performing well. Others believe that market performance is more related to the health of specific companies, and not the overall economy. One way to examine these theories is to use statistics to determine the relationship between an overall economy and its stock market. If the relationship is strong, then there would appear to be an economy-wide factor or set of factors at work. If it is weak, then a stock market's performance might be a function of specific company factors.

Table 1 Summary Statistics for Annual Return from 1975 to 1996

Country	Mean Annual Return (%)	Standard Deviation (%)
Australia	15.70	23.17
Austria	11.49	31.01
Belgium	17.42	20.13
Canada	11.82	14.29
Denmark	12.99	22.38
France	15.72	26.56
Germany	13.87	22.66
Great Britain	20.05	22.29
Hong Kong	26.36	33.47
Italy	11.10	32.82
Japan	17.20	23.77
Malaysia	16.63	26.33
Netherlands	18.87	14.70
Norway	14.48	32.54
Singapore	19.71	26.27
Spain	9.69	29.71
Sweden	18.22	21.56
Switzerland	16.17	20.22
United States	15.37	12.07

There are a number of macroeconomic variables available to test a relationship, such as inflation, gross national product (GNP), and unemployment rate. In addition, interest rates have been shown to be a key factor in explaining overall stock returns in a country or market. The main reason is relatively straightforward. When interest rates rise, the cost of borrowing for everyone, including companies, goes up. This decreased borrowing power makes it more expensive for companies to expand and grow, because it is more expensive to obtain financial resources. If growth slows, then so does the future earnings potential of companies. Therefore, the stock prices suffer as investors realize that prospects for the future are not as strong.

Because different countries have different political and economic environments, a single interest rate cannot be interpreted the same from one country to the next. For example, anything less than about 7% is considered to be a relatively low interest rate for the United States. However, in Japan, 3% is considered high. The reasoning stems from Japan's political and macroeconomic policies, which focus on keeping interest rates as low as possible to allow for as much growth as possible. In the United States, economic policy-makers prefer low interest rates to high interest rates, but do not hesitate to raise interest rates if signs of inflation appear on the horizon. These international differences suggest that care must be taken when using countries that have markedly different interest rate environments to determine the relationship between interest rates and stock market returns.

In this case we will examine the relationship between market return and the interest rate in 16 world markets in1992. Prior to investigating the strength and direction of this relationship, we conduct an **exploratory data analysis** (EDA) to examine the distribution of our variables. Histograms of market returns and interest rates in 1992 are provided in **Figure 1 and 2**. From these graphs we can observe the variability in returns and rates during this time period for this sample of markets. While these graphs show us how these variables are distributed, they do not tell us anything about how they are related, or unrelated. To examine a potential relationship between returns and rates, we use a scatter plot of returns versus rates (see **Figure 3**). From Figure 3, we can begin to observe whether there is a relationship, the direction of the relationship, and whether the relationship is linear or not.

Figure 1

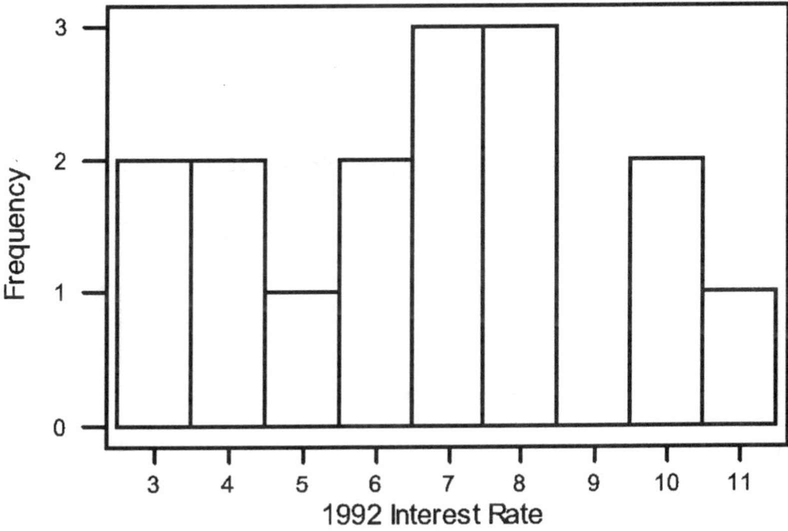

Figure 2

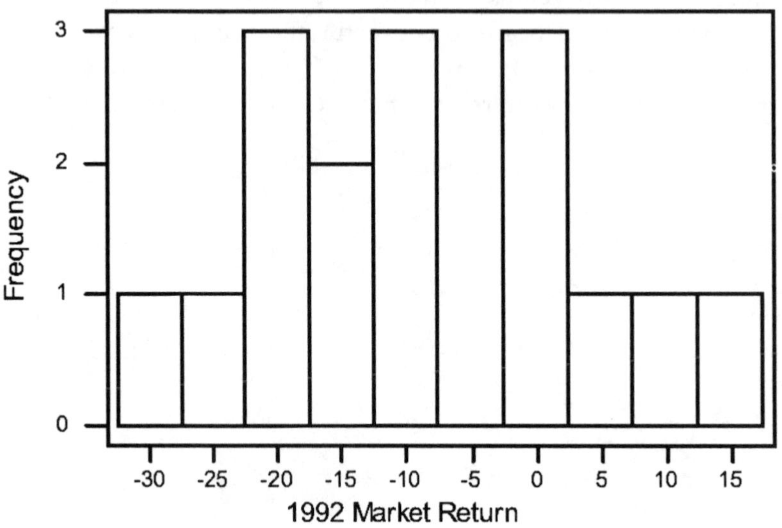

Figure 3

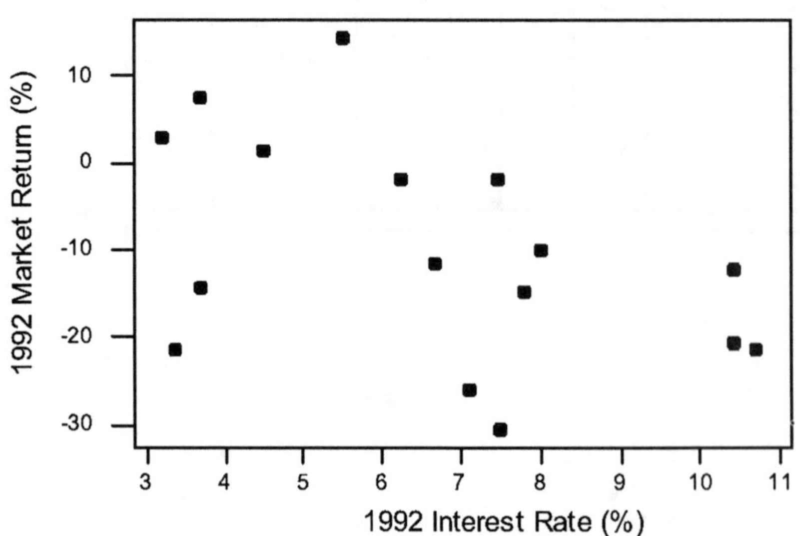

Examining Figure 3, do you suspect a relationship? Do you think this relationship will be negative or positive? (ie., As interest rates increase, do you believe that market returns decrease, or increase?) Do you think there will be any unusual observations, which perhaps should not be included in this sample of international markets? Why or why not?

Analysis Questions:

1. Note, that during the time period of 1975-1996, Japan appears to have had a better return than that for the U.S. market – thus potential for more profit for an investor. However, also note that Japan's return was more variable during this same period (greater standard deviation), and thus presented a greater risk for an investor. Using the numerical statistics provided, compute the confidence interval for the market return for both the Japanese and U.S. markets. What do these confidence intervals tell you about risk for these two markets?

2. Conduct a two-sample hypothesis test to see if indeed, there is a significant difference between the Japanese and U.S. markets during this time period. Support your conclusions with statistical evidence. If there is no significant difference in these two returns, which market would you have preferred to have your money in during this time period? What role does risk play in your decision?

3. Conduct a thorough exploratory data analysis of the return and rate data. Are market returns and interest rates for this sample approximately normally distributed? Why or why not? From your graphical analysis and numerical summaries, do you suspect any unusual observations at this point? Why or why not? From Figure 3, does there appear to be a relationship between interest rates and market return in 1992? Positive or negative? Linear or nonlinear?

4. Conduct a regression analysis of these two variables, using market return as the response variable and interest rates as the explanatory variable. Now, does there appear to be a relationship between these two variables? Interpret the coefficient of interest rates, as well as the t-statistic and p-value for this explanatory variable.

5. Evaluate the overall fit of this model. (i.e., What is R^2? How is it interpreted?) Conduct a residual analysis of this simple linear regression model. Are each of the regression assumptions satisfied? Why or why not?

6. Re-run the regression without the observation for Japan. Does it change any of the results? Why does this make economic or financial sense?

Discussion Questions:

1. What other variables or information might be important in explaining a country's overall stock return?

2. How would you summarize the relationship between these two financial variables for a novice investor in these world markets?

Source:

http://www. stanford.edu/~wfsharpe/rets/returns.htm

(Returns in this case were adapted from data provided by Independence International Associates, Inc. at this website.)

BREAK-EVEN TIME FOR NEW PRODUCTS

Statistics Topics: **Simple Regression**
Residual Analysis

Data File: **npd.xls**

To better understand new product development and to investigate the relationship between speed of product development and subsequent product sales, we collected data about new product development practices of small firms. The **sampling frame** selected for the study consisted of a wide cross-section (9 different 4-digit Standard Industrial Classification, or SIC, groups) of small manufacturing firms (less than 100 employees). Entrepreneurs (e.g., presidents or owners) were used as single key informants, since it was presumed that they had vested interest and intimate knowledge of their firms' new product development processes. Only those firms that had recently completed, or were close to the completion of the product development project, were considered for the research. Of the firms that agreed to participate in the survey, only 592 firms met the criteria of recent product development. The surveys were mailed to these 592 firms (see Exhibit 1 in Part A of this case series for the survey).

Of the 592 questionnaires mailed, 286 (48%) were returned and 246 (42.0%) were largely completed. One hundred and twenty-nine of these firms (53%) provided information about a recently completed project, whereas 114 (47%) provided information about a current product development project.

Bringing new products to market faster is critical for a start-up company to gain competitive advantage in the marketplace, as well as quick funding from venture capitalists. Many companies report impressive market performance just by reducing their product development times, but it is not known whether such results hold true for small companies. Will a faster development time lead to improved market performance for small firms?

Our main objective is to investigate whether the market performance variable of break-even time is related to product development time. One can measure initial market performance in many ways, including market share, sales, profits and time to break-even. To estimate market share, a firm needs to know the size of the served market, which is difficult to measure, especially in the early stage of the product life cycle when the market is evolving. Sales and profits are valid measures of initial market performance; however, we prefer to investigate the time from product launch to break-even since this period is influenced by market acceptance, cost of production and sales, selling price, and time value of money. Managers can use break-even time to evaluate projects across a wide variety of settings and

---

This case was prepared by Professor Abdul Ali and Professor N. R. Sharpe as a basis for class instruction and discussion. The authors acknowledge the help of Prof. Robert Krapfel and Prof. Doug LaBahn in collecting the survey data.

circumstances. Thus, time to break-even is a universally accepted and understood yardstick of market performance. The schematic diagram in **Figure 1** shows the order of occurrence of product development time and break-even time.

Figure 1

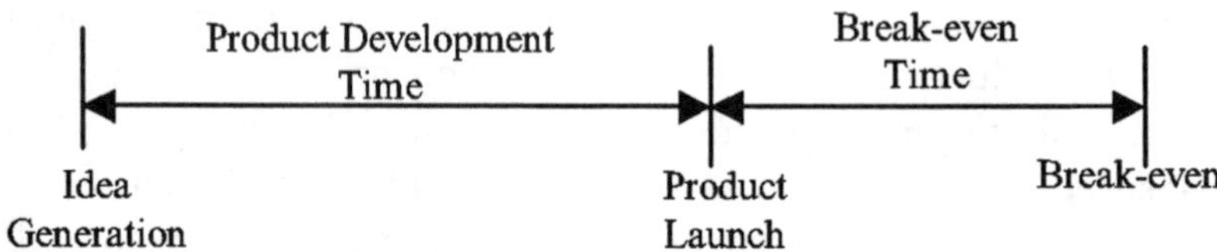

In the survey, break-even time was defined as the elapsed time from the end of product launch when the product was commercially available to the start of making profit when cumulative product contribution had repaid the development and start-up investments. Since both break-even time and development time variables are measured in months (ratio scale), they are **quantitative variables**. Thus we use a simple linear regression analysis to explore the **statistical relationship** between these two variables. We know the relationship between break-even time and development time is not **deterministic**, i.e., every time a company takes a certain time to develop a new product, it will not always break-even in the same number of months. Many other factors such as product innovation, competitive intensity in the marketplace, and launch strategy will affect the break-even time. However, we are initially interested in finding a linear equation that describes the existing relationship between break-even time and product development time.

Exploratory Analysis

First, we draw a scatter plot to examine the relationship between break-even time and development time. **Figure 2** shows this graph. We note from Figure 2 that one firm reported an unusually long development time with a short break-even time. Since this firm is not typical of our sample and may skew the results of our analysis, we decide to remove this firm from the analysis for this case. **Figure 3** below is a scatter plot of break-even time and development time showing those firms who were participants (P) and non-participants (NP) in incubator programs excluding one unusual obervation.

Figure 2

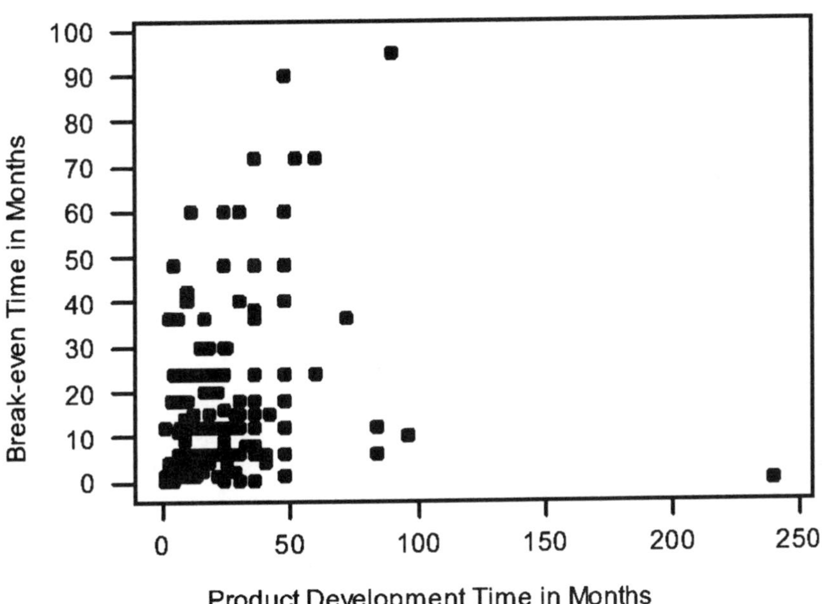

Figure 3

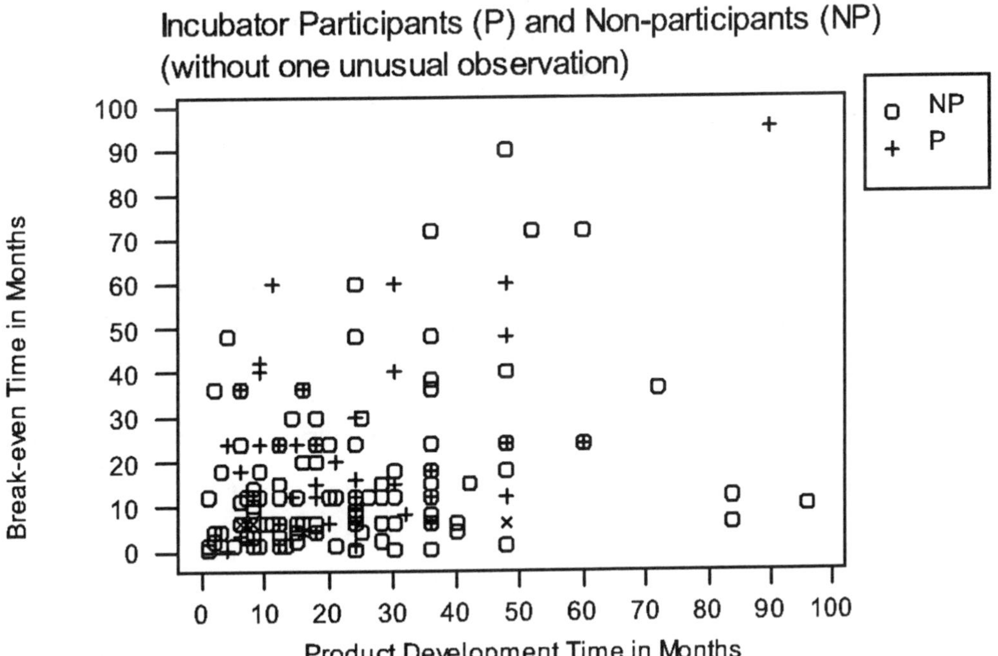

In addition, although all the firms in the sample are currently developing products, not all of these firms have *launched* these products. Since the time to break-even is measured from product launch (unlike development cost or development time), this requires a longer-range estimate for the firm. Thus we decide to use only those firms who have *completed* their product development for this analysis. **Figure 4** shows the scatter plot of these two variables with both participants and non-participants identified for only those firms who have launched their products.

Figure 4

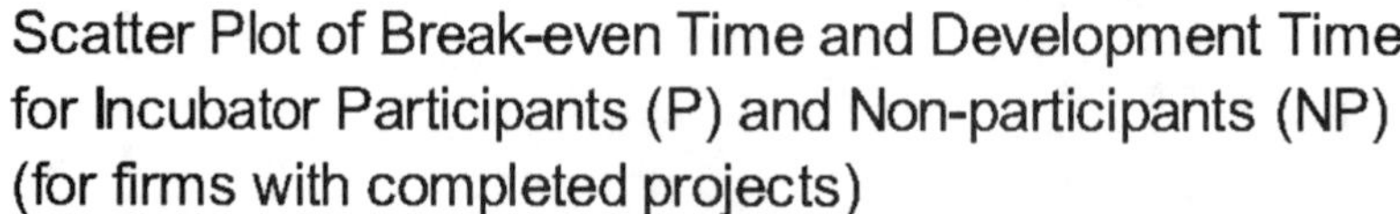

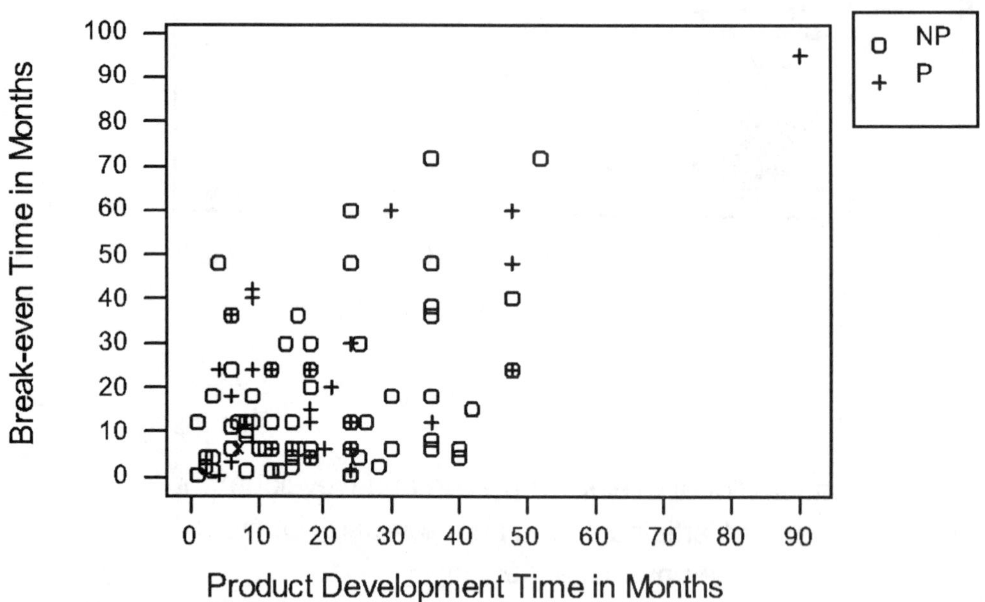

What do you observe from Figure 4? Do you believe that a relationship exists between these two project outcome variables? Do you suspect that this relationship is positive – that a longer development time will place a firm at a disadvantage by resulting in a longer period until breaking even on the product? Do you suspect a linear relationship? Do you believe that the relationship will be different or the same for the two types of firms: participants and non-participants in incubator programs?

Regression Analysis

A hypothesized relationship between break-even time and development time can be represented by the following model:

$$BT = \beta_0 + \beta_1 \, DT + \varepsilon \qquad (1)$$

where, BT = Break-even time,
DT = Product development time,
ε = Error term.

The regression output for this linear model is shown below in **Table 1**.

Table 1

Regression Analysis

```
The regression equation is
BT = 6.11 + 0.656 DT

96 cases used 24 cases contain missing values

Predictor        Coef       StDev          T        P
Constant        6.110       2.683       2.28    0.025
C1             0.6558      0.1099       5.97    0.000

S = 15.94       R-Sq = 27.5%     R-Sq(adj) = 26.7%

Analysis of Variance

Source           DF          SS          MS         F         P
Regression        1      9049.8      9049.8     35.63     0.000
Residual Error   94     23872.9       254.0
Total            95     32922.7
```

As Table 1 suggests, R^2 is approximately 27%, and the coefficient of development time is significant in the model ($p<0.05$) – indicating that firms with faster product development reported shorter break-even times. However, we should also analyze the residuals for this simple linear regression to investigate 1) the presence of outliers; 2)the normality of the residuals; and 3) the relationship (if any) between the residuals and the predicted values of break-even time. **Figure 5** below contains the residual plots for the simple linear regression model presented in Table 1.

Figure 5

Residual Plots for Simple Regression Model

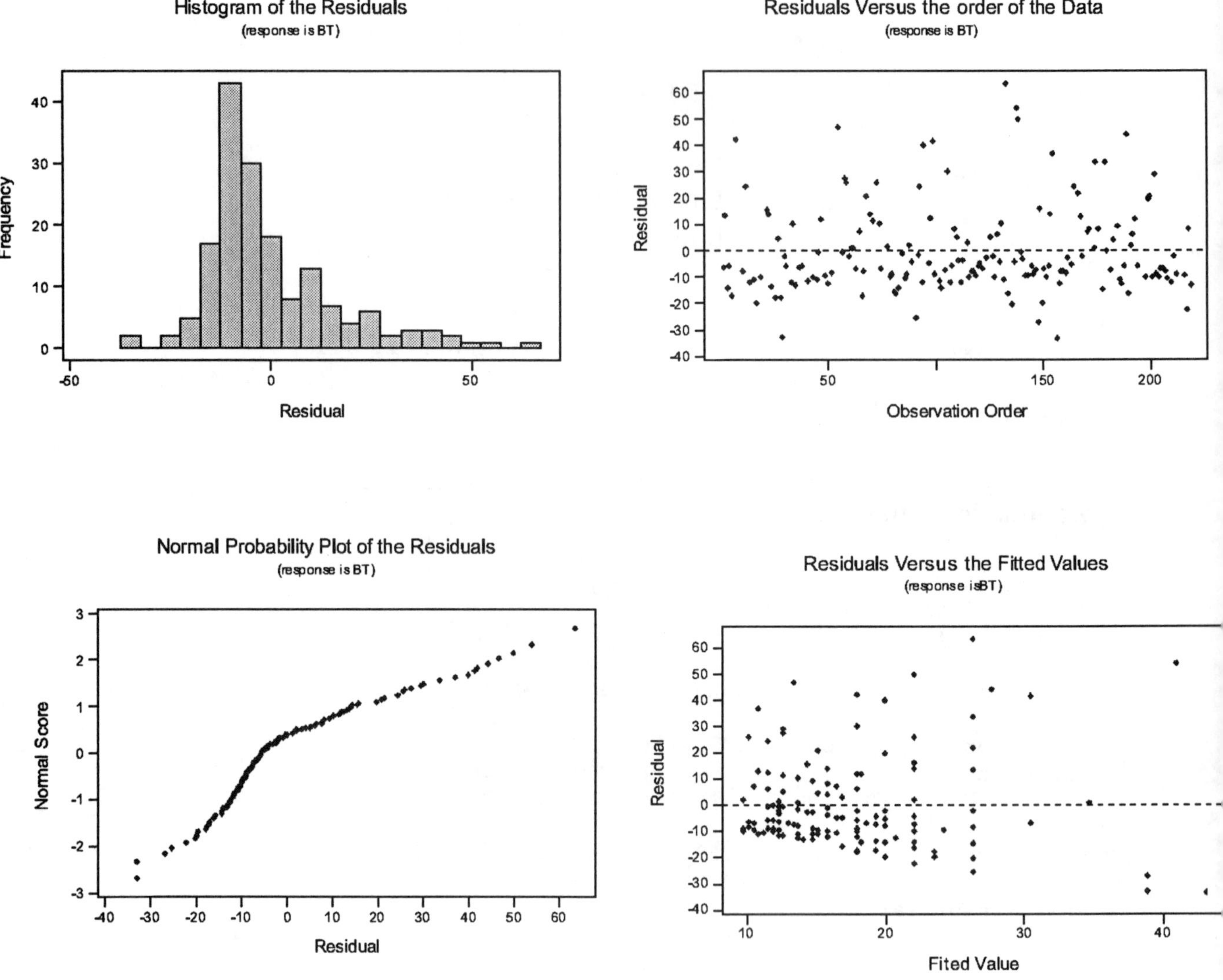

Summary

The analysis in this case supports the general theory among marketing professionals that a shorter development time for a new product can lead to a shorter break-even time – faster to market means a more immediate impact on the bottom line. This theory has been well-established for large firms, but not specifically for small firms. Thus our analysis confirms the significant positive relationship between faster new product development time and break-even time for small firms. However, although this relationship is significant, clearly there other factors that must influence the break-even time of small firms (R^2 is only 27%).

Does participation in an incubator program have an impact on the length of time a firm takes to break-even with its new product? As we have seen in Part (C) of this case series, those firms who launched their products and participated in incubator programs actually have a *longer* (although not significantly longer) break-even time than non-participants. Shall we then run separate regression analyses for incubator participants and non-participants to describe the relationship between break-even time and development time? Why do we expect incubator participants to be different from non-participants in managing their new product development process? Are these incubator participants smaller in size compared to non-participants? Does this mean incubator participants lack the marketing expertise necessary to lunch a new product? Do incubator participants have as much experience in developing new products as non-participants do? If we do find such differences, shall we then conclude that participation in an incubator program may help a firm, who has lack of experience and resources in developing a new product, but may not provide much help in launching the new product? What do you think? We ask you to investigate this plausible theory below in the case analysis.

Conceptual Questions:

1. Interpret the coefficient of development time in Table 1. (i.e., If development time is decreased by 1 month, how much can break-even time be expected to change?)

2. From Figure 4 what can you conclude about the residuals for this simple linear regression model? Are the standard assumptions for linear regression satisfied?

Analysis Questions:

1. Develop, using the data set, a regression model for break-even time using development time and participation in an incubator program. Use a dummy variable to represent incubator participation.

Let, $X_2 = 0$ if the company did not participate in an incubator program and $X_2 = 1$ if the company did participate in an incubator program.

Your regression equation will now be:

$$\hat{Y} = b_0 + b_1X_1 + b_2X_2,$$

Where,

b_1 = slope of development time
b_2 = incremental effect of incubator participation on break-even time.

2. Explain why the slope for development time will be the same in both models and why only the Y-intercept will change. Does participation in an incubator program have an impact on break-even time? Is your coefficient of determination (R^2) higher for this model, than in the earlier simple linear regression model?

3. (Optional: Multiple Regression)
   Now add another variable representing the size of the firm (D4) to the regression model you have developed. Is this new variable significant in the model? What is the effect on the overall fit of the model?

Discussion Questions:

4. From your analysis above, what type of relationship appears to exist between development time and break-even time? Is this relationship different for incubator participants and non-participants? Is this relationship different when the size of the firm is held constant? What other factors do you think might impact break-even time for small firms?

MUTUAL FUND FLOWS (A)

Statistics Topics: **Simple Linear Regression**
Correlation
Residual Analysis
Dummy Variables

Data File: **fundflows.xls**

Investing in the stock market has become a national hobby for many. In fact, there are now billions of dollars invested in the United States stock market through mutual funds every month. The level of investment activity has not always been this high. Fifteen years ago, money flowing into mutual funds was consistently measured in the millions, rather than billions, and there were many months where there was more money being taken out of mutual funds than being put into mutual funds.

Growth in the money flowing into mutual funds has ranged from steady to extreme. During the entire calendar year 1984, approximately $5.9 billion of new money was invested in 471 different stock mutual funds. Today, there are over 3,000 stock mutual funds, and money flowing into these funds can reach over $40 billion, during a single month!

Experts have suggested many reasons for this growth in mutual fund investing. These reasons include the wealth of information investors have at their disposal, the ease of investing an entire portfolio with a single mutual fund company, and being able to invest in a lot of companies with relatively few dollars. However, the reason most cited for the proliferation of mutual funds has simply been the performance of the United States stock market. In other words, because the stock market has been doing so well, more investors want a "piece of the pie" and prefer to have an "expert" manage their portfolios, rather than management their own investments. In addition, it is believed that when the market performs poorly, investors pull their money out of the market and put it elsewhere.

The Investment Company Institute (ICI), which is the trade group for mutual fund companies such as Fidelity, Vanguard, and Templeton, tracks mutual fund activity. The mutual fund companies report to ICI facts and figures each month, including mutual fund performance and money flowing into and out of mutual funds. The Institute then aggregates the figures and releases them to the general public through the press and their web site. These numbers are closely watched by many investors and public officials, who see mutual fund flows as a measure of the psychology of the stock market.

Table 1 below reports US stock market performance and money flowing into stock mutual funds on an annual basis from 1984 through 1996.

Table 1

Annual Stock Market Performance and Fund Flows		
Year	**Stock Market Return (%)**	**Fund Flows (in millions)**
1984	7.26	$ 5,873.9
1985	28.64	$ 8,454.5
1986	17.90	$ 21,724.2
1987	8.62	$ 19,042.3
1988	16.15	$ -16,107.9
1989	27.85	$ 5,789.2
1990	-1.59	$ 12,810.8
1991	28.47	$ 39,438.6
1992	7.53	$ 78,949.4
1993	9.23	$ 129,397.3
1994	1.50	$ 118,948.1
1995	32.69	$ 127,586.7
1996	21.41	$ 216,873.9

From the table above, it does not appear that the two variables are related. However, note that following 1995 (a year of extremely high performance for the market), there was a jump in investment in the market. In addition, note from the data, that the investment in the market appears to have steadily grown in the 1990's, as opposed to the 1980's, when the fund flows were more volatile, and in fact, appear to have varied more with the annual market returns.

Before coming to a conclusion on the relationship between fund flows and stock returns, you may decide to take a closer look at the data. Your intuition might be telling you that investors probably follow the stock market more than once a year when making their investment decisions. Thus we might want to use data that are measured more frequently. **Figure 1** below represents a scatter plot of fund flows versus market returns on a ***monthly*** basis.

Table 2 contains the output from Minitab for the **simple linear regression** model using market returns to explain monthly fund flows. Note that the R-Square is relatively low (4.2%), although the coefficient of market return is significant ($p < .05$). This lack of meaningful explanatory power indicates that either a transformation might be helpful, or other factors should be considered to explain the variation in fund flows. Since the distribution of fund flows appear to be slightly skewed (more positive flows than negative), a logarithmic transformation might be useful to normalize the distribution of the response variable. However, it can be shown that a **log-linear model** only increases the R-square to 7%. Thus what other approaches or techniques can we use to improve the model of the relationship between fund flows and market return?

Figure 1

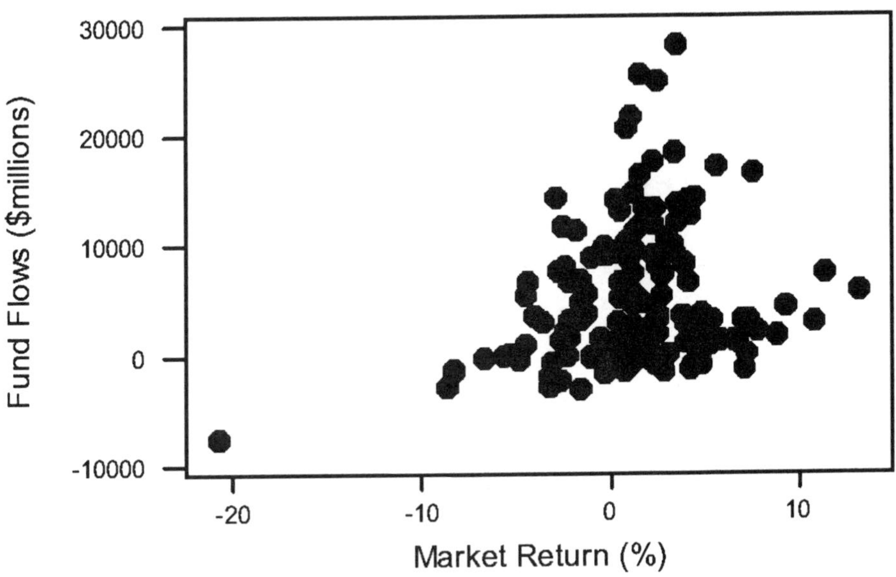

Table 2

Regression Analysis

The regression equation is
Fund Flows ($millions) = 4589 + 310 Market Return (%)

Predictor	Coef	StDev	T	P
Constant	4589.1	523.3	8.77	0.000
Market R	310.2	121.1	2.56	0.011

S = 6145 R-Sq = 4.2% R-Sq(adj) = 3.5%

Analysis of Variance

Source	DF	SS	MS	F	P
Regression	1	247867348	247867348	6.56	0.011
Residual Error	151	5701218835	37756416		
Total	152	5949086183			

Let's examine the relationship between fund flows and market return more closely. Remember that these are observations collected over two decades. In fact, when we observed the original annual data, we noticed that the magnitude of investment in mutual funds was markedly greater after 1990. Suppose we create a graph where we can visually see whether the observations are from after 1990 or not. **Figure 2** is a scatter plot of fund flows versus market return, which displays the decade of each observation.

Figure 2

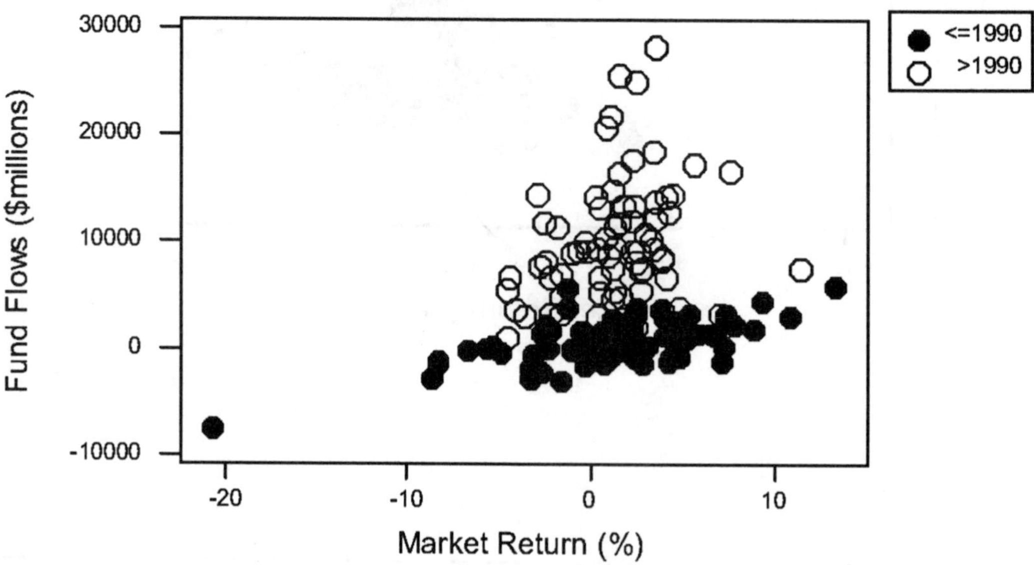

Note the difference in both magnitude and variability between the two decades. How can we take into account the noticeable difference in the magnitude of investment in the market across the two decades? If we create a new variable to represent the two decades (e.g., 'on or before 1990' and 'after 1990'), this will be a categorical variable. How can we represent a categorical variable in our model? We will use a **dummy variable** to represent our two categories:

$X_2 = 0$ if the observation is on or before 1990 and
$X_2 = 1$ if the observation is after 1990.

The regression equation now appears as:

$$\hat{Y} = b_0 + b_1X_1 + b_2X_2,$$

where now

b_1 = slope of market return with fund flows, holding constant the decade
b_2 = incremental effect of the decade of observation while holding constant the effect of market return.

Note that this model assumes the slope of market return with fund flows is the same for the two decades and that only the intercept will be different. This is because when the observation is from the 1980's, the second term will be zero, resulting in the equation:

$$\hat{Y} = b_0 + b_1X_1.$$

When the observation is after 1990, then X_2 will be equal to 1 and the resulting equation will be:

$$\hat{Y} = (b_0 + b_2) + b_1X_1.$$

Thus, the slope of market return in both models is b_1. **Table 3** is the regression model including the dummy variable for the decade of observation.

Table 3

Regression Analysis

The regression equation is
Fund Flows ($millions) = 288 + 300 Market Return (%) + 9170 Decade

Predictor	Coef	StDev	T	P
Constant	288.1	465.6	0.62	0.537
Market R	299.73	80.39	3.73	0.000
Decade	9169.9	660.8	13.88	0.000

S = 4079 R-Sq = 58.0% R-Sq(adj) = 57.5%

Analysis of Variance

Source	DF	SS	MS	F	P
Regression	2	3452752468	1726376234	103.73	0.000
Residual Error	150	2496333714	16642225		
Total	152	5949086183			

Note that the coefficients for both market return and decade are significant ($p < .05$) and R-Square is now 58%. The coefficient of market return can be interpreted as suggesting that a 1% increase in the stock market will result in a $300 million increase in fund flows if we hold constant the decade of the observation. The coefficient of our dummy variable (Decade) suggests that in the 1990's investors put an average of $9.17 billion more into mutual funds than in the 1980's. **Figure 3** shows the resulting models for the two decades.

Figure 3

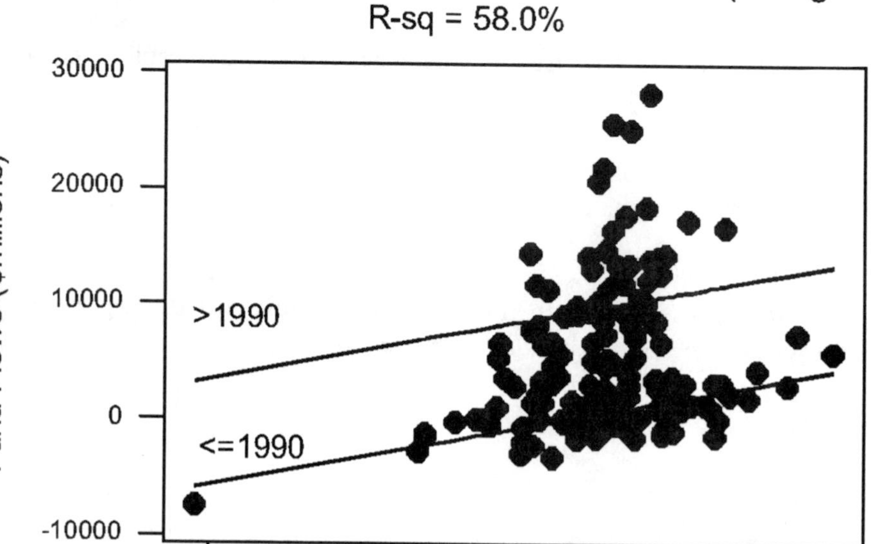

While this current model now seems to explain the relationship between fund flows and market return fairly well, we still have not addressed the increased volatility in investment in mutual funds in the 1990's. In addition, should the models for the two decades really have the same slope? To explore this last question, let's develop models for each decade separately. **Figure 4** below is the regression model for the 1980's and **Figure 5** contains the residual plots for this model.

Note, that the R-square for this model using only data from 1984-1990 is now almost 40%. The same model for the data during the period 1991-1996 produces an R-square of only approximately 5% and a coefficient for market return that is not significant ($p > .05$). Why is there a stronger relationship between investment in mutual funds and market performance during the 80's than the 90's? Several reasons have been suggested to explain this. Most notably, the demographics of the average mutual fund investor have changed. Investors during the 1990s are younger and less established in their careers, but have fewer long-term financial obligations. As such, they are more apt to take more risks and move their investments around more often. This behavior has resulted in a much less consistent investment strategy than the fund investors of the 1980s. Adding to this was a proliferation of additional information investors could use to determine investment strategy, as well as a growth in marketing efforts by mutual fund companies. Therefore, many more factors now impact the flow of money in and out of mutual funds than during the 1980s.

Figure 4

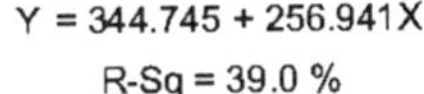

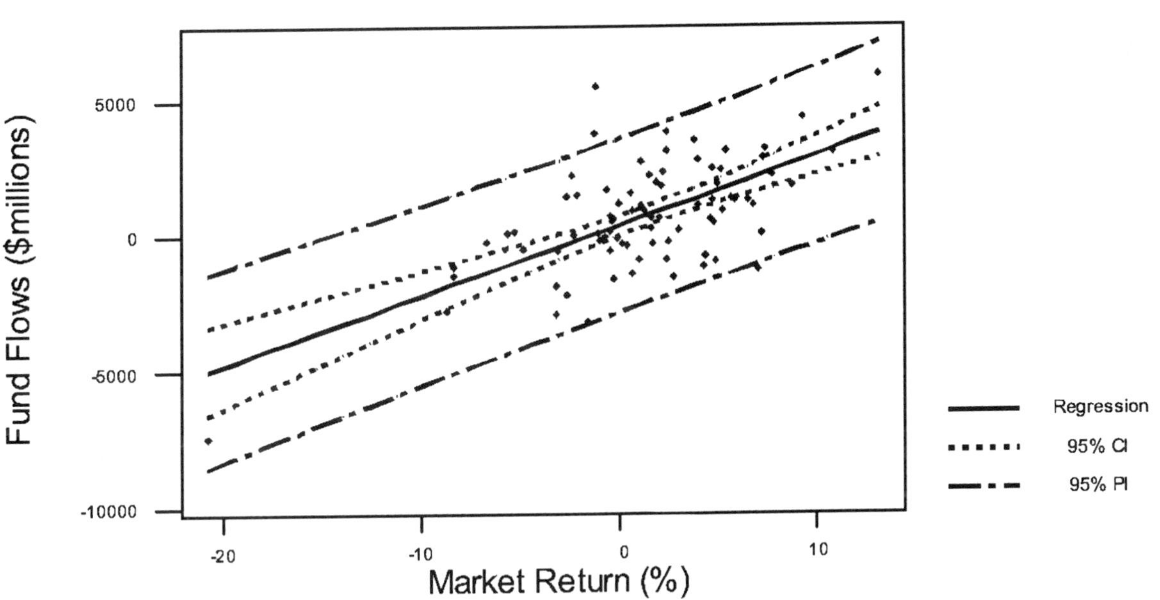

Figure 5

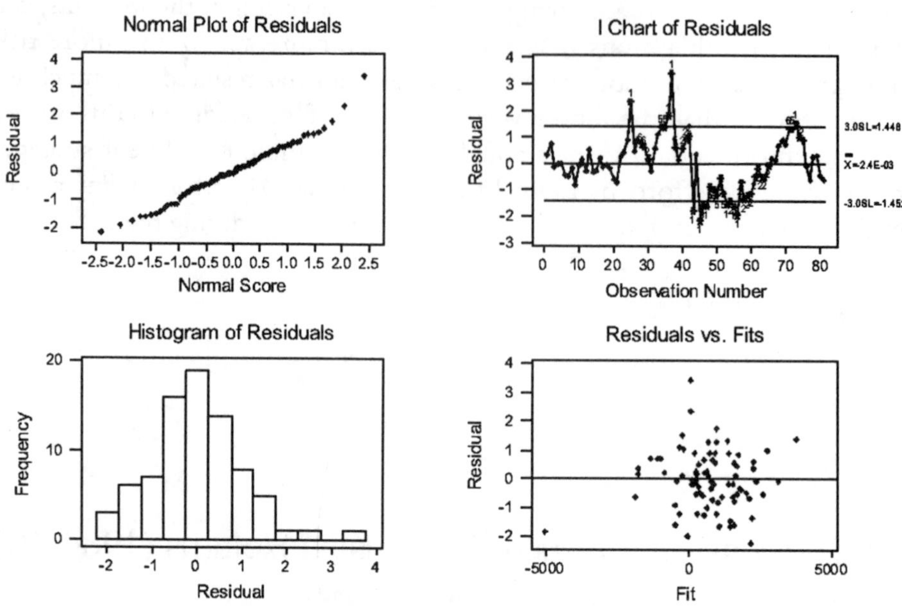

Conceptual Questions:

1. What is the difference between the 95% **confidence interval** and 95% **prediction interval** in Figure 4? Why is the prediction interval wider than the confidence interval?

2. Examine the residual plots in Figure 5. Are the residuals approximately normally distributed for this model? Do the residuals appear to be random over time, or do they appear to have a pattern? Do any regression assumptions appear to have been violated?

Analysis Questions:

1. Using the regression output below for separate models of the 1980's and 1990's, compute the 95% confidence intervals for the slopes of market return using the formula:

$$CI = b_i \pm t_{n-2}(StDev_{bi}).$$

   Interpret the meaning of these confidence intervals. Are the two slopes for these two models significantly different?

Regression Analysis for 1984-1990

The regression equation is
fundflows80 = 345 + 257 return80

Predictor	Coef	StDev	T	P
Constant	344.7	184.4	1.87	0.065
return80	256.94	36.12	7.11	0.000

S = 1603 R-Sq = 39.0% R-Sq(adj) = 38.3%

Regression Analysis for 1991-1996

The regression equation is
fundflows90 = 9264 + 439 return90

Predictor	Coef	StDev	T	P
Constant	9263.5	746.5	12.41	0.000
return90	438.6	231.6	1.89	0.062

S = 5705   R-Sq = 4.9%   R-Sq(adj) = 3.5%

Discussion Questions:

1. Given the lack of a relationship between market performance and investment in mutual funds in the 1990's, what other factors do you think might help explain fund flows now?

2. How might each of the following groups of people be interested in the results of your analysis?
   a. Mutual fund managers
   b. Investors
   c. Government officials (who worry about stock market volatility)
   d. Managers of companies whose investors are now large mutual funds rather than individual investors

---

Sources:

http://www.stanford.edu/~wfsharpe/rets/returns.htm
(Returns in this case adapted from data provided by Independence International Associates, Inc., found at this website.)

http://www.ici.org
(Fund flows data obtained from this website.)

BASEBALL STADIUM AGE AND ATTENDANCE

Statistics Topics: **Multiple Regression**
Dummy Variables

Data File: **attendance.xls**

Baseball is the national pastime, and each year millions of spectators go to the major league ballparks to cheer on their favorite teams, root for their favorite players, and enjoy one of the oldest sports in America. When the big-money players are in town, fans flock to the stadium to see them in action. Furthermore, attendance is usually higher when the home-team is winning a majority of their games, as loyal supporters gather to see the local "boys of summer" in pursuit of a World Series appearance.

Over the past decade, owners have argued for bigger and better stadiums. Two new ballparks are scheduled to open in 2000: Comerica Park for the Detroit Tigers costing an estimated $240 million and Pacific Bell Park for the San Francisco Giants with an estimated cost of $305 million. In addition, new ballparks are planned for the Pittsburgh Pirates in 2001 ($211 million), the San Diego Padres in 2002 ($270 million), and the Cincinnati Reds in 2003 ($231 million). Currently, a hot debate is broiling in Boston over the price tag of a new Fenway Park for the Red Sox – with an estimated cost of $350 million. Boston fans and residents are skeptical of the need for such expense.

What is fueling this race for new stadiums? One of the arguments of team owners is that a modern stadium will help to attract the "star" ball players and hence enhance their ability to field a stronger team. They argue that a newer and larger stadium – with it's additional club seats and luxury boxes – will attract more fans, thus generating more revenue, which can be used to attract the top players in the league, and increase team performance. The league estimates that the net increase in revenue of new ballparks is approximately $29 million a year per team. But what's in it for the local fans? Are the new stadium amenities and state revenues raised from additional out-of-state fans worth the increase in ticket and vending prices? Will a new stadium actually impact team performance? Is the age of the stadium or team performance related to attendance at a ballgame?

Some baseball clubs continue to set attendance records each season, even when the team does worse than the year before. The reverse also holds true: a team may improve in the league standings, but stadium attendance decreases. In this case we investigate the relationship between baseball stadium attendance, age of a ballpark, and team performance, as measured by the team's winning percentage (number of games won divided by the total number of games played).

Conceptual Questions:

1. What type of relationship do you think may exist between stadium attendance and team winning percentage?

2. Do you think that the age of the ballpark would have any relationship with stadium attendance? Do you think that average attendance per game is the correct variable to use to measure attendance? Why or why not? If not, please suggest an alternative measure using the variables in the data set. Form a hypothesis for the relationship between age of ballpark and stadium attendance.

Analysis Questions:

1. Create new variables for your analysis. Use the variables Capacity and 1998 Average Attendance to create a new "percent capacity variable." Use the year the ballpark was built to create a variable representing "age" of ballpark.

2. Conduct an exploratory data analysis (EDA) of the data. Investigate the distribution of each of the variables (Average Attendance in 1998, Age of Ballpark, and Winning Percentage in 1997 and 1998) and the relationship among these variables using graphs and a correlation matrix. What kind of relationship do you suspect among these variables?

3. Conduct a regression analysis for these variables using the percent capacity as the dependent variable. Evaluate the fit of the model. Are the coefficients significant? How good is the fit? Should any variables be removed?

4. Do you think that the "age of the ballpark" is the correct variable to use in the regression? Create a dummy variable for age as follows:

   X = 0 if the ballpark was built before 1987 and
   X = 1 if the ballpark was built in or after 1987.

   Rerun the regression using the new dummy variable for age of ballpark. Should any variables be removed? Are there unusual observations? Evaluate the fit of the model.

Discussion Questions:

1. State your conclusions regarding this analysis. Can you conclude that prior team performance and/or age of ballpark influence percent of capacity that is filled at a ballpark? Does this conversely imply that a "newer" ballpark will enhance attendance and/or team performance? How does collinearity between age of ballpark and 1997 winning percentage impact your conclusions? What are confounding factors that you have not considered in your analysis?

Sources:

Vaillancourt, M. *Fresh Parks are proving a Boon for Many Teams,* Boston Globe, Monday, May 17, 1999.

Responsive Database Services, Inc., *Amusement Business,* 110(43), October 26, 1998.

MUTUAL FUND FLOWS (B)

Statistics Topics: **Multiple Regression**
Multicollinearity

Data File: **fundflows.xls**

In part A of this case, you examined the relationship between fund flows and the stock market's return during the period 1984 to 1996. The financial theory states that the size of investment in the market is directly related to the performance of the market, as measured by return; if the market performs poorly, investors pull their money out, while if the market performs better than expected, investors jump on the bandwagon and invest heavily in the market. During the period of your sample, while the relationship was *statistically* significant, the *strength and meaning* of the relationship was less clear. (Recall that the simple linear regression between the two variables produced a significant slope, but an R^2 of only approximately 4%, without the dummy variable for decade.)

Why might this be the case? If you think about it, investors probably consider more than just how the stock market is doing when deciding whether or not to put money into it. For example, what if people didn't have any money to invest? Obviously, this would impact the flows into stock mutual funds. What if you could earn a competitive rate of return on a savings account? You may put your money into a bank rather than the stock market. Finally, what if you were really worried about the impact of inflation eroding the future purchasing power of your income? In times of extreme economic downturns, investors resort to hard assets such as gold, silver, and copper, which tend to retain their value even if the stock market takes a dive.

In this case, you are provided with several additional economic variables: Gold prices, interest rates on 1-year certificates of deposits (CDs), and U.S. disposable per capita income. The gold prices are provided per ounce, and fluctuate heavily over the time period. Of course, if you're investing in gold you want to buy low and sell high, just as if you're investing in stocks. Or, you want gold to retain its value better than other investments on a relative basis. Normally, if you're investing in stocks, you're investing for longer than a few days (except for the notorious "day-traders"). As such, the appropriate interest rate you could earn on your funds would not be the rate on a savings account, but rather on a longer investment such as a certificate of deposit, as they tend to pay higher rates of return than savings accounts. The certificates, or CDs, are like loans to the bank. You write the bank a check and they lend out the money to someone. After a year, they pay you back, along with interest. Unlike traditional savings accounts, however, you are basically locked in for the whole time period, and have to pay a penalty to withdraw early. But, as mentioned before you earn a higher rate of interest. To get a measure of how much money people have to spend after taxes, a measure of disposable income is used. Since these funds can be spent on both essentials, as well as savings, only a portion can be used to invest.

Analysis Questions:

1. Conduct an Exploratory Data Analysis of the data in this case. (i.e., First, examine the distribution of each of the variables graphically – Market Returns, CD Interest Rates, US Disposable Income, and Gold Prices.) Are they approximately normal? Skewed?

2. Continue your exploratory analysis. Examine the relationship between each of the variables using scatter plots. (In Minitab, it may be efficient to use the menu function Graph > MatrixPlot.) What do you observe? Which variables do you think are positively related to Fund Flows? Negatively related to Fund Flows? Most strongly related to Fund Flows? Do you think the relationships are linear? If not, why not?

3. Examine the correlation between each of the variables using a correlation matrix. (Create the matrix using the Fund Flows variable in the first row and column.) Which variables are positively correlated with Fund Flows? Negatively related to Fund Flows? Most strongly related to Fund Flows? Are any of the other variables related to each other? Do you suspect multicollinearity? Support your answers with interpretation of the statistical results in the correlation matrix. Do your answers make sense in the context of financial theory?

4. Conduct the regression analysis using Fund Flows as the response (dependent) variable and the other financial variables as the explanatory (independent) variables. Examine the coefficients. Do they agree with the results of your exploratory data analysis conducted above? Why or why not? Are all of the variables significantly related to Fund Flows during this time period? Explain. Interpret the meaning of the coefficient (slope) of CD Interest Rates in the context of this model.

5. Evaluate the fit of your model. Examine the measures of adjusted R^2, standard error of the estimate, the multiple F-test, and individual t-tests for each of the slopes. Discuss any unusual observations in your data set. Are they potential outliers (high residuals) or leverage values (unusual 'X' values)? Provide statistical evidence. Do you think any of them should be removed? Why or why not?

6. Evaluate the validity of the regression assumptions using residual plots. Is each of the assumptions met? Why or why not?

Discussion Questions:

1. Summarize the relationship among these financial variables. Would you recommend that this model be used to make investment decisions? If not, suggest at least one way to improve the model (e.g., specific transformations, other variables, etc.).

Sources:

http://www.stanford.edu/~wfsharpe/rets/returns.htm (Market returns obtained from this website.)

http://www.ici.org (Fund flows obtained from this website.)

Datastream International (U.S. income, interest rates, and gold prices obtained from this source.)

MUTUAL FUND FLOWS (C)

Statistics Topics: **Multiple Regression**
Autocorrelation
Lagged Variables

Data File: **fundflows.xls**

In part A of this case, you examined the relationship between fund flows and the stock market's return during the period 1984 to 1996. To develop an improved model of the pattern in fund flows over time, in part (B) of this case, you added various U.S. economic indicators: interest rates on 1-year certificates of deposits (CDs), U.S. disposable per capita income, and gold prices per ounce. In this case, you will be asked to treat the data as a **time series** and examine the relationship over time among the errors, or residuals.

Suppose that based on your analysis in the previous cases, you arrive at a model using Market Return, U.S. Disposable Income, CD Interest Rates, and CD Interest Rates squared for fund flows over the period 1984-1996.

Table 1

Regression Analysis

```
The regression equation is
Fund Flows ($millions) = - 54365 + 358 Market Return (%)- 4108 CD_Int_Rate +
                         4.22 US_Disposable_Income_per_Capita + 227 CD-squared

Predictor        Coef       StDev          T        P        VIF
Constant       -54365        9995      -5.44    0.000
Market R       357.65       75.59       4.73    0.000        1.0
CD_Int_R      -4108.0       649.1      -6.33    0.000       21.8
US_Dispo       4.2219      0.5319       7.94    0.000        1.6
CD-squar       226.70       46.44       4.88    0.000       22.5

S = 3827        R-Sq = 63.6%     R-Sq(adj) = 62.6%

Analysis of Variance

Source            DF          SS          MS         F        P
Regression         4  3781883939   945470985     64.57    0.000
Residual Error   148  2167202244    14643258
Total            152  5949086183

Durbin-Watson statistic = 0.51
```

Below is the graph of residuals over time for the multiple regression model above:

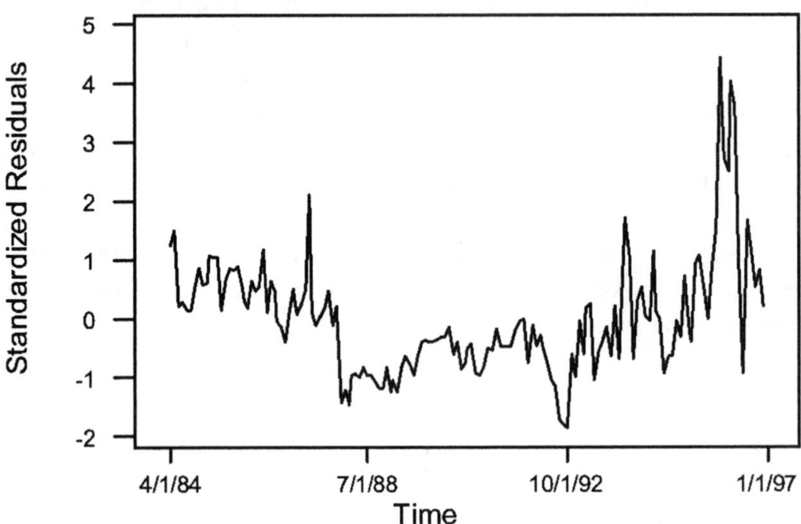

Figure 1

In Figure 1, do you observe a pattern in the residuals over time? This graph can be used to examine if the errors are independent of each other over time. If there exists a pattern in this graph, we say that adjacent residuals are related to each other over time and that **autocorrelation** exists. If adjacent residuals are related, a "cyclical" pattern ususally can be observed in the graph of residuals over time.

Observations of cyclical patterns in such residual plots can often be difficult. However, we can test for the presence of autocorrelation using the **Durbin-Watson statistic**. The Durbin-Watson statistic is defined as:

$$D = \frac{\sum (e_i - e_{i-1})^2}{\sum e_i^{\,2}},$$

where e_i = error, or residual, at time period i.

The Durbin-Watson statistic (D) can be obtained as part of the standard regression output in most statistical packages. In Table 1, the Durbin-Watson statistic appears at the bottom of the table and equals 0.51. How do we interpret this number? Note, that if the residuals are perfectly correlated (equal to each other), then the numerator of D equals zero. If the residuals are negatively correlated (equal, but opposite in sign), then the numerator of D equals 4 $\sum e_i^2$ and D equals 4. Thus we would like to know if D is between 0 and 4, or sufficiently close to 2, for autocorrelation to not be an issue. How can we determine "how close is close enough?" Just as for the F-statistic, our answer depends on the sample size (n) and the number of predictors in the model (p). Also, as for the F-statistic, we need to obtain critical values. For D, we have two critical values: the lower and the upper critical value. The decision rules based on these critical values appear below:

If $D < D_L$ (lower critical value), then positive autocorrelation exists
If $D_L < D < D_U$, then we are not sure if correlation exists
If $D > D_{U,}$ then positive autocorrelation does not exist.

These lower and upper critical values are obtained from a table of values derived using a significance level of .05 for various values of n and p.

How can we address the issue of autocorrelation? When adjacent errors are related, it usually means that future values of the response variable are related to prior values of the response variable, as well as prior values of the predictor variables.

Does this make sense in the context of our financial data? Why would prior investment in mutual funds affect future investment in funds? One reason could be that investment in mutual funds has become part of an individual's portfolio or savings plan for the future, and therefore monthly investment remains fairly stable. In addition, money flows into mutual funds through retirement plans (administered by companies or employers) that tend to remain stable over the course of a year. In the same vein, why would prior market returns affect future fund flows? It is well known that investors follow performance of stocks over time prior to investing. Therefore, it is intuitive that prior market return would impact an investor's decision regarding *when* and *how much* to invest in the market. Finally, the monthly U.S. disposable income, which is a measure of investors *ability* to invest in the market, is cumulative in its impact on the depth of investors' pockets; therefore, prior disposable income might impact the flow of money into mutual funds.

How can we represent prior values in a regression model? Since our current model considers predictions at time t, we would like to include observations at time t-1 to reflect prior observations. Thus our model looks like:

$$\hat{Y}_t = b_o + b_1Y_{t-1} + b_2X_t + b_3X_{t-1} ,$$

where Y_{t-1} = observation for response variable at time t-1
X_{t-1} = observation for predictor variable at time t-1.

These variables representing observations at prior time periods are called **lagged variables** and are easily created in most statistical packages. If the variables represent observations at time t-1, we say that they have been lagged once, if they represent observations at time t-2, then we say that they have been lagged twice, and so on.

Below is the multiple regression model including Market Returns, U.S. Disposable Income, CD Interest Rates, and CD Interest Rates-squared, as well as the lagged variables representing each of these variables at time t-1.

Table 2

Regression Analysis

```
The regression equation is
Fund Flows ($millions) = -12131 +367 Market Return(%) -1288 CD_Interest_Rate
                  -1.60 US_Dis_Income + 53.6 CD-squared + 0.778 lagfundflow
                  - 207 lagreturn + 297 lagCDRate + 2.56 lagincome

Predictor        Coef       StDev          T        P        VIF
Constant       -12131        7017      -1.73    0.086
Market R       367.38       47.31       7.76    0.000        1.0
CD_Inter      -1287.8       708.1      -1.82    0.071       65.6
US_Dispo       -1.597       1.247      -1.28    0.202       22.1
CD-squar        53.65       31.34       1.71    0.089       25.7
lagfundf      0.77769     0.05142      15.12    0.000        2.8
lagretur      -206.58       51.01      -4.05    0.000        1.2
lagCDRat        296.5       525.0       0.56    0.573       36.9
lagincom        2.557       1.227       2.08    0.039       21.6

S = 2378         R-Sq = 86.4%      R-Sq(adj) = 85.6%

Analysis of Variance
Source            DF          SS          MS          F         P
Regression         8  5125406910   640675864     113.32     0.000
Residual Error   143   808456729     5653544
Total            151  5933863639

Source         DF       Seq SS
Market R        1    247315390
CD_Inter        1   2433024027
US_Dispo        1    781052705
CD-squar        1    327411253
lagfundf        1   1221423838
lagretur        1     88358170
lagCDRat        1      2266059
lagincom        1     24555467

Durbin-Watson statistic = 2.19
```

Conceptual Questions:

1. Are each of the economic indicators in Table 2 significantly related to investment in mutual funds over the period 1984-1996? Evaluate the fit of the model. Are you still concerned with autocorrelation? Evaluate the presence of multicollinearity in the model. Would you remove any variables? If so, which variables would you remove? (Note, if you are not familiar with the **Variance Inflation Factor** (VIF), then you may use a correlation matrix to investigate collinearity.)

2. Why do think that both a linear term *and* a quadratic term for CD Interest Rates are included in the above model? How would you have discovered this nonlinear relationship in your Exploratory Data Analysis (EDA)?

Analysis Questions:

1. Use the Best Subsets Procedure to evaluate different models using different combinations of the variables included as predictors in Table 2. Which model would you recommend that a financial analyst use to predict fund flows based on other economic indicators (past or present)? Why?

2. Develop a regression model for the combination of variables you recommend as a result of your best subsets procedure. Evaluate the fit of your model. Examine the residuals over time for the model you have selected. Do you observe a pattern in the residuals over time?

3. Does your model fit equally well over the entire time frame of your sample (i.e., 1984-1996)? If not, why not? (You may use information from parts A and B of this case.)

Discussion Questions:

1. How would you summarize the relationship among these financial variables? Would you recommend that this model be used to make investment decisions?

Sources:

http://www.stanford.edu/~wfsharpe/rets/returns.htm (Market returns obtained from this website.)

http://www.ici.org (Fund flows obtained from this website.)

Datastream International (U.S. Income, interest rates, and gold prices obtained from this source.

SALES IN A SEASONAL INDUSTRY

Statistics Topics: **Time Series**
Moving Averages
Exponential Smoothing
Seasonal Modeling

Data File: **coolpool.xls**

Cool Pool & Spa, Inc. (CPS) constructed and maintained residential swimming pools in the area around Pearl River, New York, an affluent suburb of New York City. The business had provided Armando Insignares and his family a good living since he had bought the company in 1978. It now supported five full-time and 20 seasonal employees. Armando credited much of his success to being able to manage the ups and downs of this business. Inventories of chemicals, building supplies, and accessories had to be just right or he could be stuck with them for an entire season, and he had to have the right number of part-time employees lined up for the summer, or he wouldn't be able to fill demand.

In 1997, his son Mike, a junior at Babson College, realized that tools he was learning might be able to help his father develop more accurate inventory and staffing strategies. When he arrived home for Christmas vacation, he gathered sales information from July 1994 through November 1997. He wanted to see what these data suggested for the future success of the business. He realized, however, that he would have to account for the margin of error in his sales estimates and that if his efforts were to be valued, these forecasts would have to be used.

Perhaps the most difficult chore would not be forecasting, but convincing his father of the model's value. He described his father as a swarthy, 48 year-old, who sported a full beard; a hands-on worker who had learned the business from the ground up and had always operated by having a "good pulse" on the business.

History of the Company

Armando Insignares was born in 1950 in Colombia and had immigrated to New York City with his family in 1957. His father died in 1961, leaving him responsible for helping support the family. He kept a series of part-time jobs while he attended school in Hackensack, New Jersey. In 1968, he went on to college where he met his future wife, Irene. Finances were tight, so two and a half years later he left college to earn money and support his mother and younger brothers and sisters.

In 1978, he heard that CPS was for sale from Irene's boss and used a combination of savings and loans to consummate the deal. Although he knew nothing of the business, he was accustomed to manual labor and liked working outside.

This case was prepared by Professor Norean Radke Sharpe with the assistance of Heather Miller, Research Assistant, and David Wylie, Director of Case Development, Babson College, as a basis for class instruction and discussion.

The core of the CPS business was from warranty work maintaining and repairing Sylvan pools. Armando severed his ties with Sylvan in 1993, however, so that he could start constructing pools on his own, and hopefully increase his revenue stream.

Operations

CPS obtained most of its work either from the yellow pages, word-of-mouth, or from advertisements in the local paper during the late spring and summer. Construction was usually planned in the spring, although orders often came as late as mid-July. Once the season, started the pace of construction was good. In fact, in the last few years, Armando had to turn down a number of new pools because he could not get to them during the season. Repair work could come at any time, but peaked early in the year when pools were first opened and again in the latter part of the summer when pools had received frequent use. Since everyone seemed to want to open their pools at the same time under their yearly contracts, repair work could be scheduled more easily than new construction.

CPS hired seasonal staff, who focused on opening pools under the yearly contracts, with the first crunch just before Memorial Day weekend. Then half the staff devoted itself to construction, and the other half was almost evenly split between ongoing maintenance (pool cleaning, pump maintenance, etc.) and repair. Armando bought some materials and supplies in bulk, particularly for those things that were used or sold every year, such as chemicals or paint. He received a discount of about 20% if he ordered early in the spring, but most of these supplies did not keep well over the winter, so any left over at the end of the season would have to be discarded.

Sales Forecasting

Mike knew that if he could help his father predict sales then his father would be able to strategize the future of his business. He would have a better idea of when hire seasonal employees and how to manage his inventory. Mike had learned about several different forecasting techniques from his quantitative courses at Babson. He decided to view the data graphically to see which methods might be appropriate (see **Figure 1**). The data represent sales from July 1994 through November 1997 (see **Table 1**).

After graphing the sales figures, Mike noticed a seasonal pattern in the data, which was not surprising, since his father's business was highly seasonal. He also observed that over this period of time, average annual sales remained fairly constant. Thus he realized he was dealing with a stationary time series.

Mike wondered which methods of forecasting he should use given the seasonal nature of the data. He knew there were several available and he also wondered how he should choose between them. How would he know which was the better model? The most important concern was that the model chosen not only accurately model historical sales, but also accurately predict the company's future sales. Mike thought he could use the sales figures from 1994 to 1997 to predict 1998 sales (see **Table 2**) and compare the alternative models for accuracy in forecasting future sales.

Figure 1

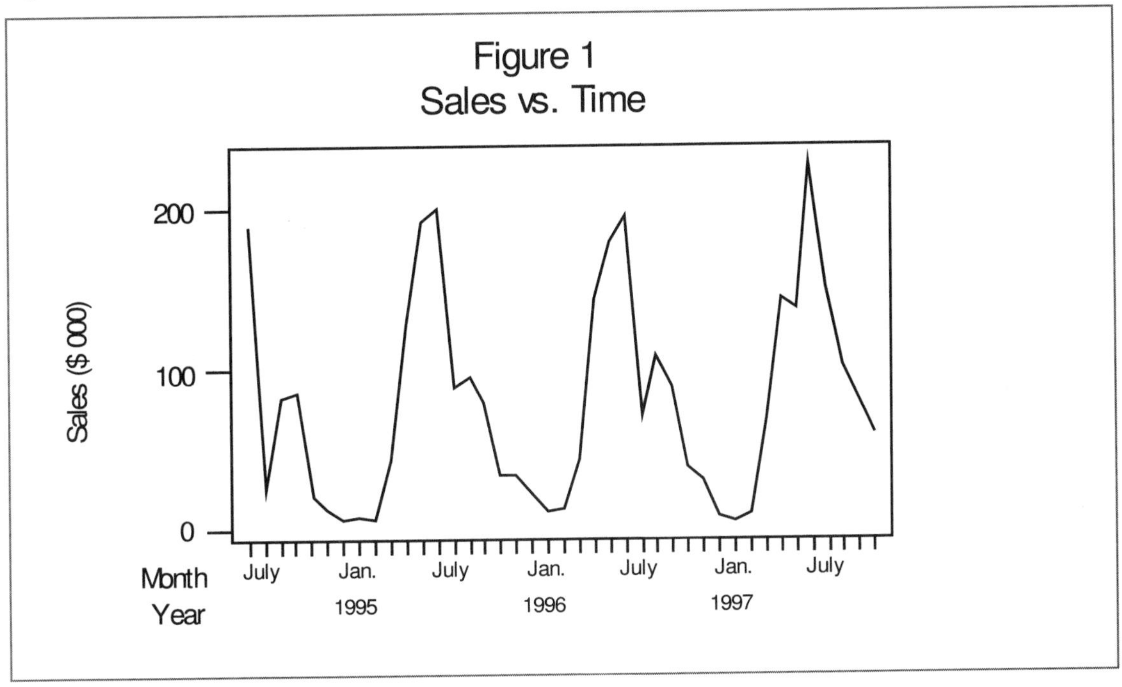

Table 1

Historical Sales Data

Period	Month	Sales	Period	Month	Sales	Period	Month	Sales
1	Jul-94	$ 189,511	15	Sep-95	$ 95,077	29	Nov-96	$ 37,969
2	Aug-94	$ 24,617	16	Oct-95	$ 79,550	30	Dec-96	$ 30,043
3	Sep-94	$ 81,409	17	Nov-95	$ 33,000	31	Jan-97	$ 7,986
4	Oct-94	$ 85,148	18	Dec-95	$ 32,710	32	Feb-97	$ 4,532
5	Nov-94	$ 20,040	19	Jan-96	$ 21,691	33	Mar-97	$ 8,773
6	Dec-94	$ 12,145	20	Feb-96	$ 11,576	34	Apr-97	$ 67,655
7	Jan-95	$ 6,500	21	Mar-96	$ 12,389	35	May-97	$ 144,127
8	Feb-95	$ 7,200	22	Apr-96	$ 42,987	36	Jun-97	$ 136,471
9	Mar-95	$,732	23	May-96	$ 142,892	37	Jul-97	$ 227,415
10	Apr-95	$ 43,835	24	Jun-96	$ 179,529	38	Aug-97	$ 149,200
11	May-95	$ 27,743	25	Jul-96	$ 195,498	39	Sep-97	$ 100,830
12	Jun-95	$ 192,618	26	Aug-96	$ 70,144	40	Oct-97	$ 81,021
13	Jul-95	$ 199,482	27	Sep-96	$ 107,770	41	Nov-97	$ 59,261
14	Aug-95	$ 88,357	28	Oct-96	$ 88,334			

Table 2

New Sales Data		
		Sales
43	Jan-98	$ 5,065
44	Feb-98	$ 19,602
45	Mar-98	$ 19,956
46	Apr-98	$ 84,697
47	May-98	$ 158,695
48	Jun-98	$ 245,221

Analysis Questions:

1. Using the sales in 1994 through 1997 forecast the sales for the first six months in 1998. What are the advantages and disadvantages of using **Moving Averages** to obtain forecasts? of using **Exponential Smoothing** to obtain forecasts?

2. How can you model the seasonality? Use any approaches that you have learned in your course (e.g., **Seasonal Dummy Variables** in Multiple Regression or **Seasonal Decomposition**) to obtain forecasts for the first six months in 1998.

3. **Optional:** Use the partial autocorrelation function to investigate the presence of autocorrelation in the time series. Develop an appropriate autoregressive model to forecast sales for the first six months of 1998.

Discussion Questions:

1. Compare the forecasts from the alternative models that you have developed with the actual sales in 1998. Which model should Mike suggest his father use? Should he use different models during peak and off-peak seasons? Why or why not?

2. What advice regarding staffing, inventory, and marketing should Mike give his father? Assume that Mike is going to play a major role in CPS after graduating from Babson and intends to computerize the entire operation. What should Mike measure and track over time to improve the accuracy of his model?

Source:

The data for this case were obtained with permission from the owner of Cool Pool & Spa, Inc.

BOSTON SUNDAY GLOBE (A)

Statistics Topics: **Time Series**
Seasonal Decomposition

The Globe Newspaper Company is located in Boston, Massachusetts. On Monday through Saturday, the company publishes the daily **Boston Globe**, commonly known in and around Boston as "the Globe," and on Sunday it publishes the **Boston Sunday Globe**, or simply "the Sunday Globe." Each edition of the Globe is comprised of the "A" section, with coverage of major local and regional events as well as global and national issues and separate sections for New England and for "Metro/Region" news items. In addition to these general news sections, there is an extensive sports section, an arts section, classifieds, and, depending on the day of the week, other special interest sections, such as Automotive, Food, Health/Science, etc. The Sunday Globe has extended coverage of most of the above listed areas and includes extra pull out pieces such as a book section, a television weekly, **Parade** magazine, and the **Boston Sunday Globe Magazine** which includes special articles on local celebrities, an expanded crossword puzzle, etc. Therefore these newspapers appeal to a market that extends well beyond the city of Boston and its metropolitan area. Both the Globe and the Sunday Globe enjoy the largest circulation (average daily unit sales) among all comparable daily or Sunday newspapers in New England.

The Boston Globe was founded in 1872. One year later Gen. Charles H. Taylor became its first publisher. Gen. Taylor's family subsequently ran the newspaper for 120 years. In 1993 The New York Times Company, publishers of **The New York Times** ("the Times"), acquired Affiliated Publications, Inc., the then parent company of the Globe, for $1.04 billion. At the time of the acquisition the Globe was the nation's thirteenth largest newspaper based upon daily circulation.

The stated intent of the New York Times Company in 1993 was to allow the current Globe management team to retain editorial control of the newspaper. "We have no intention of going into Boston and telling them how to run The Boston Globe," claimed Arthur Ochs Sulzberger, the New York Times Company chairman and chief executive. William O. "Bill" Taylor, great-grandson of Gen. Charles H. Taylor and chairman and president of Affiliated Publications, was named chairman and chief executive of Globe Newspaper Company. He also continued to serve as publisher of the Globe, the position he had held since 1978.

On April 1, 1997, Benjamin B. Taylor, Bill Taylor's second cousin, became only the fifth publisher of the Globe in its 125 year history. However, on July 12, 1999, the New York Times Company named Richard H. Gilman, senior vice president of operations of The

---

This case was prepared by Professor Steven Eriksen, as a basis for class instruction and discussion.

New York Times, publisher of The Boston Globe effective immediately. This abrupt movement ended the 126 year era of a Taylor family member in the position of publisher of The Boston Globe.In the late 1990's The New York Times Company hired The Strategic Pricing Group, a marketing consulting company located in Marlborough, Massachusetts, that specializes in pricing policy, to conduct a statistical analysis of historical circulation data of the Times. Some of the results of this study seemed quite remarkable to managers at the Times, and so they had The Strategic Pricing Group consultants present their findings to Globe management in Boston. Consequently the Globe decided to conduct a similar analysis of their circulation data.

The objectives of this study at the Globe were:

- To identify the factors that appear to impact upon circulation based upon historical data
- To develop a mathematical model of the relationship between circulation and these variables
- To address such questions as how will changes in controllable and uncontrollable variables impact upon circulation

The first phase of this study focused upon The Sunday Globe. The data used in this phase consisted of time series of circulation (average number of Sunday copies sold in the month), product related variables (price per copy, number of home delivery contractors, number of papers returned unsold, etc.), and demographic/economic variables (area population, unemployment rates, etc.). Each series consisted of monthly data from January 1987 to March 1998 (n = 135 observations).

Circulation figures were separated into two groups, **Home Delivery** and **Single Copy**. Home delivery circulation is the number of customers who subscribe to The Sunday Globe. The subscription price typically involves a discount from the "newsstand price," or the single copy price published at the top of the front page of the paper. New subscribers are often attracted by promotions offering, albeit temporarily, very deep discounts from the newsstand price. Single copy circulation consists of purchases at newsstands, college bookstores, supermarkets, etc., and from the ubiquitous green vending boxes scattered all about the Greater Boston area. While home delivery customers rarely pay the full newsstand price, single copy purchasers practically always pay the newsstand price.

Two approaches to data analysis were utilized in this study, **time series analysis** and **multiple regression analysis**. The time series analysis addresses questions such as:

- How have the levels of a single variable (such as circulation) changed over time?
- Does there appear to be a predictable pattern?

The multiple regression analysis addresses questions such as:

- Can we identify the relationship between the **response variable** (ex.: circulation) and one or more **explanatory variables** (ex.: price, economy related variables)?
- What inferences can we draw from this relationship regarding management decision making (ex.: pricing policy)?

The time series plot for home delivery circulation is shown in **Figure 1**. (This figure and all others in this case were generated by **Minitab**.)

Figure 1

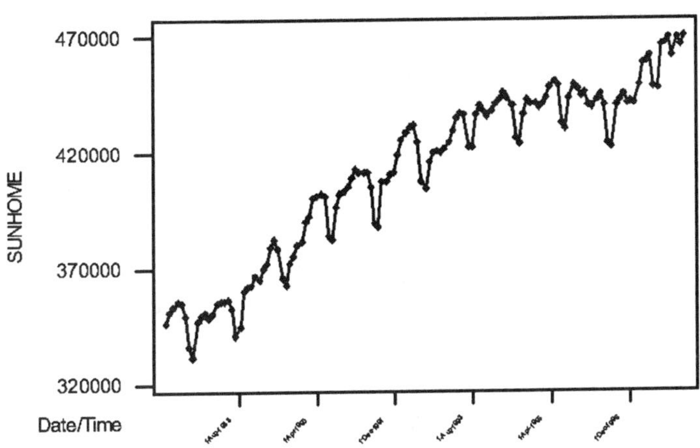

Figure 2 is a Minitab graphic that contains four plots. The plot in the upper left corner is the same as Figure 1. The Detrended Data plot consists of the residuals from a **linear trend line** fitted through the original series. The Seasonally Adjusted Data is the **deseasonalized** series, meaning the raw data with the seasonality taken out, or equivalently the original circulation values divided by their corresponding **seasonal indices** (in this case monthly indices). The plot in the lower right corner is the residuals after the linear trend and the seasonal effects have been removed.

Figure 2

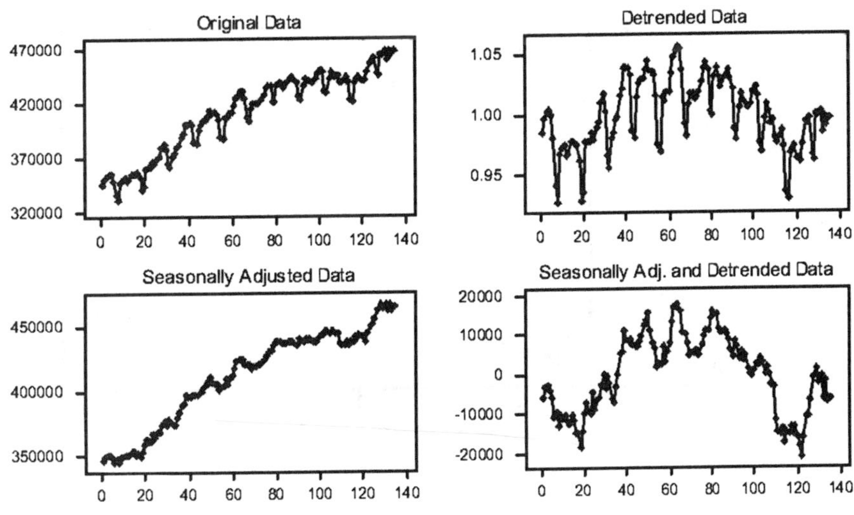

Figure 3 is a Minitab graphic that also contains four plots. The plot in the upper left corner is a chart of the seasonal (monthly) indices.

Figure 3

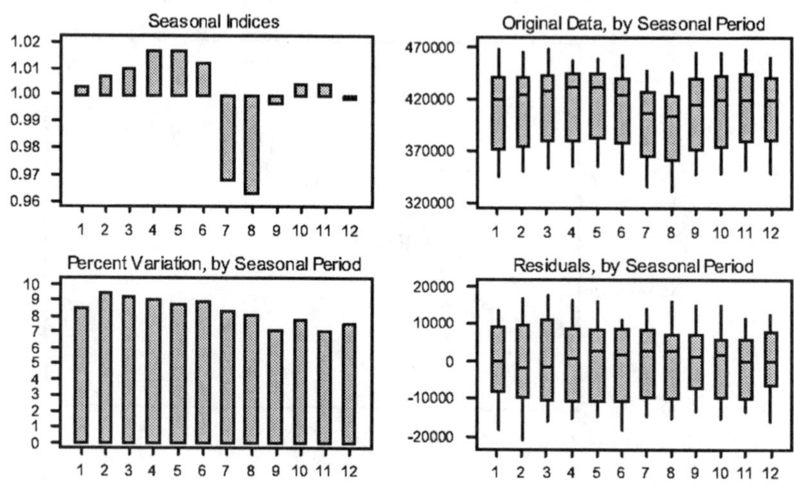

It is a widely held belief within the publishing industry that a newspaper, particularly a Sunday paper, is a **cyclical product**. When the economy is favorable, so the theory states, circulation will rise creating greater sales revenue, thereby driving up advertising revenue. Similarly, when the economy is in recession newspaper sales will soften resulting in declines in both sales and advertising revenue to the publishers. The unemployment rates in Massachusetts for the time frame under study are provided in a time series plot in **Figure 4**.

Figure 4

Time Series Plot of Unemployment Rate

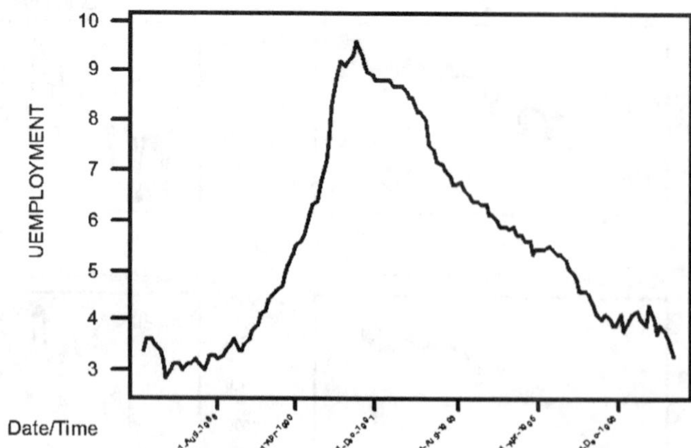

Conceptual Questions:

Examine the time series plot of circulation in Figure 1:

1. Is there an apparent trend in the data? If so, is this trend positive or negative? What implications do these answers provide about the success over these eleven plus years of home delivery sales of the Sunday Globe?

2. Is the trend linear or nonlinear?

3. Do there appear to be any recurring seasonal (monthly) effects? If so, which months appear to be the peak or the lull months? What economic factors do you suppose contribute to this predictable pattern?

Examine the graphs in Figure 2:

4. Does the information provided by the Detrended Data plot confirm your answer to Question 3?

5. Does the information provided by the Seasonally Adj. And Detrended Data plot confirm your answer to Question 2?

Examine the graphs in Figure 3:

6. Does the information provided by the Seasonal Indices plot confirm your answers to Questions 3 and 4?

Examine the time series plot of unemployment in Figure 4:

7. Describe the pattern of the unemployment rates in Massachusetts over this period of time. What does this series tell us about the economy in Massachusetts from the late 1980's to the late 1990's?

8. Does this series seem to correspond (correlate?) in any way to any of the series observed in Figure 2?

9. Does any of this information tend to support the adage within the publishing industry that a newspaper is a cyclical product?

Discussion Questions:

1. Do you regularly read a Sunday newspaper? If so, what are your reasons for so doing?

2. Do you regard the purpose of reading a Sunday newspaper to be purely informational?

3. Is reading a Sunday newspaper an expensive or inexpensive activity?

BOSTON SUNDAY GLOBE (B)

Statistics Topics: **Time Series**
Seasonal Decomposition

Single Copy sales of the **Boston Sunday Globe** refer to newspapers that are sold at retail, such as in stores, from vending machines, or from on-the-street hawkers. In **Boston Sunday Globe (A)** we analyzed the historical **Home Delivery** circulation, or subscription sales, of the Sunday Globe. Now in part (B) we examine historical single copy circulation data.

The time series plot for single copy circulation is shown in **Figure 1**. The time series consists of the average number of Sunday copies sold in each of the 135 months between January 1987 through March 1988 (same time frame as in the home delivery analysis).

Figure 1

Time Series Plot of Sunday Single Copy Circulation

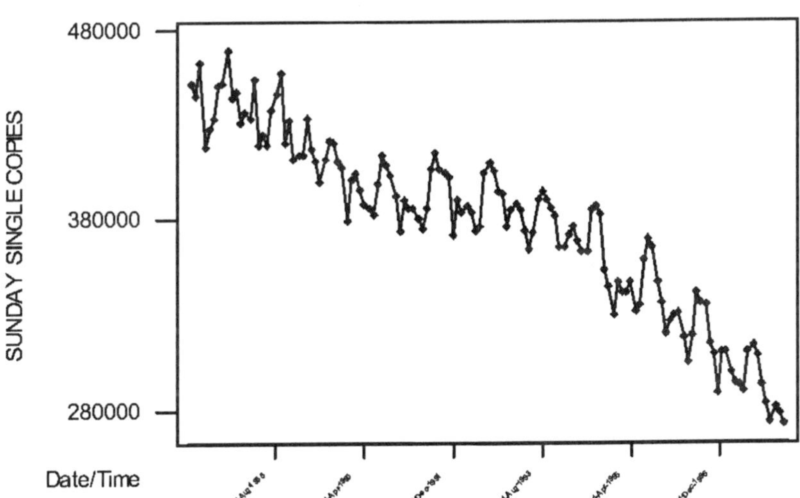

Figure 2 contains four graphs. The upper left plot is the original series for Sundays single copy circulation (same as Figure 1). The remaining three plots show the raw series with a **linear trend** removed, the **seasonality** removed, and both the linear trend and the seasonality removed.

Figure 2

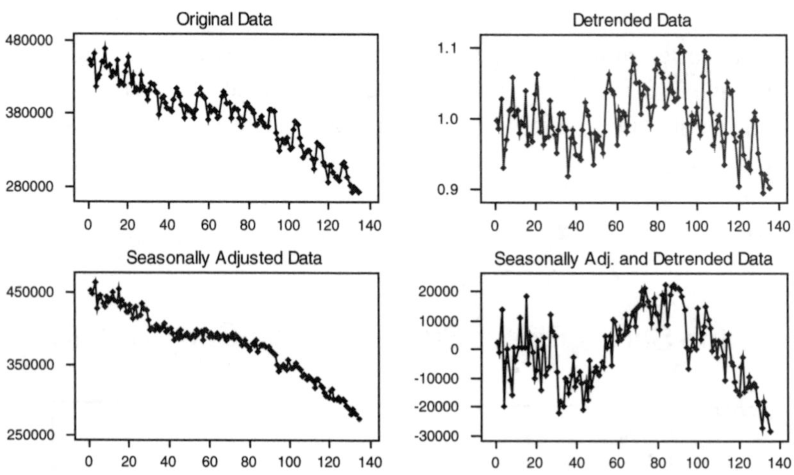

Figure 3 contains graphs for single copy circulation on Sunday, which provide information on the seasonal variation and the seasonal indices produced by a seasonal decomposition model. The plot in the upper left corner is a chart of the seasonal (monthly) indices and the plot in the lower left corner is a chart of percent variation by month.

Figure 3

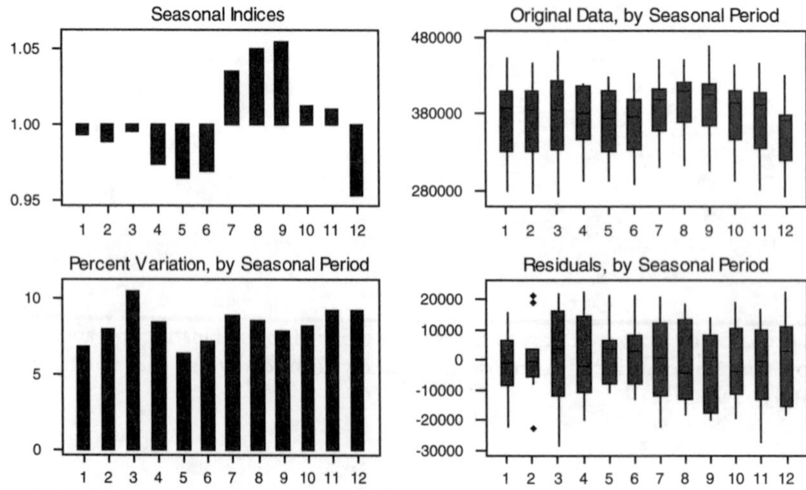

Conceptual Questions:

Examine the time series plot in Figure 1:

1. Is there an apparent trend in the data? If so, is this trend positive or negative? What implications do these answers provide about the success of single copy sales of the Sunday Globe during this time?

2. Is the trend linear or nonlinear?

3. Do there appear to be any recurring seasonal (monthly) effects?

Examine the graphs in Figure 2:

4. Does the information provided by the Detrended Data plot confirm your answer to Question 3?

5. Does the information provided by the Seasonally Adj. and Detrended Data plot confirm your answer to Question 2?

6. In Question 8 of Boston Sunday Globe (A) the series of unemployment rates in Massachusetts was compared to the series in Figure 2 of that case. Does this unemployment rate series (Figure 4 in Boston Sunday Globe (A)) correspond in any way to any of the series observed in Figure 2 in this (B) case?

7. Does single copy circulation of the Sunday Globe appear to be a **cyclical product** (strong sales in robust economy, soft sales in recession), an **anticyclical** product (strong sales in recession, soft sales in good economy), or neither?

Examine the graphs in Figure 3:

8. Does the information provided by the Seasonal Indices plot confirm your answers to Questions 3 and 4?

9. Which months appear to be the peak or the lull months?

10. If average Sunday readership is reasonably constant from month to month, one could argue that the seasonality of home delivery circulation and the seasonality of single copy circulation would be mirror images. The peak months for home delivery would be the lull months for single copy, and vice versa. Compare the answer to Question 9 with the answer to the corresponding item in Question 3 of Boston Sunday Globe (A). Does the seasonal pattern of home delivery and single copy circulation appear to be a mirror image of each other?

11. What economic factors do you suppose contribute to the seasonal pattern of single copy sales?

Discussion Question:

1. The seasonal effects that were observed in the analysis of single copy circulation were based upon historical data. Is there reason to believe that, due to developments in economic and/or technological factors, that this pattern may change in the foreseeable future?

BOSTON SUNDAY GLOBE (C)

Statistics Topics: **Multiple Regression**
Logarithmic Transformations

The study of historical circulation data undertaken by management of the Globe Newspaper Company used two statistical approaches to analyze the data: **time series analysis** and **multiple regression analysis.** In **Boston Sunday Globe (A)** we applied time series analysis to uncover information about trends and seasonal patterns of **home delivery** circulation. In **Boston Sunday Globe (B)** a similar analysis was conducted on historical **single copy** sales data.

Multiple regression analysis was applied in an attempt to determine what factors, both controllable (e.g., price) and uncontrollable (e.g., economic or demographic shifts) impact circulation based upon historical data. Many models attempting to determine the relationship between the **response variable**, circulation, and a set of **explanatory variables** were developed in the Globe study. These model specifications differed in a variety of ways. For example, in some models different subsets of explanatory variables were used. In other specifications some of the variables were **transformed** (e.g., logarithms of variables were used). We now apply the "best" set of regression models, in terms of statistical integrity, to address questions, such as: how will changes in controllable and uncontrollable variables impact circulation?

Our first model is an estimated relationship between home delivery circulation and four explanatory variables. The second is an estimated relationship between single copy circulation and three explanatory variables. The home delivery regression equation involves a **linear** specification. The home delivery model is a **multiplicative**, or a **log-linear,** specification.

A linear model is an equation in which the relationship between the response variable and the explanatory variables is assumed to be linear. Equation (1) is the expression for the hypothetical linear model.

$$Y = \beta_0 + \beta_1 X_1 + \beta_2 X_2 + \ldots + \beta_k X_k + \varepsilon, \qquad (1)$$

where Y is the **response variable**
$X_1, X_2, \ldots, X_k$ are the **explanatory variables**
$\beta_0, \beta_1, \beta_2, \ldots, \beta_k$ are the **model parameters**
ε is the **error term**

and k is the number of explanatory variables.

This case was prepared by Professor Steven Eriksen, as a basis for class instruction and discussion.

The hypothetical model corresponding to a multiplicative model, or logarithmic transformation, is shown below in equation (2).

$$Y = \beta_0 X_1^{\beta_1} X_2^{\beta_2} \ldots X_k^{\beta_k} \varepsilon \quad (2)$$

The specification in equation (2) is also referred to as "log-linear." When we take the logarithms of both sides of this nonlinear equation, the model becomes linear with respect to the **logarithms** of the original variables, as shown in equation (3) below.

$$\log Y = \log \beta_0 + \beta_1 \log X_1 + \beta_2 \log X_2 + \ldots + \beta_k \log X_k + \log \varepsilon \quad (3)$$

There are a number of reasons why a data analyst would select the multiplicative specification (2) over the linear specification (1). Three of these motivations are:

- An underlying theory from economics may dictate that the variables are related in such a manner.
- The data analyst notices while looking at scatter plots, that the relationships between the response variable and the individual explanatory variables do not appear to be in straight lines, as expected with a linear specification. When the analyst transforms the variables by taking their logarithms and then plots the logarithm of Y vs. the logarithms of the individual X variables, the relationships appear to be much closer to linear.
- The estimates of β_1, β_2, etc. in the log-linear model are estimates of the **elasticity** of the response variable with respect to the individual explanatory variables. Therefore, the estimate of β_1 in equation (3) is the expected percentage change in Y when the variable X_1 is increased by one percent, and all other X variables are held constant.

The Minitab regression output for one of the models that was generated to analyze the structural relationship between home delivery sales of the Sunday Globe and a set of explanatory variables is presented in **Figure 1**. The variables used in this equation are defined as follows:

DSUNHOME: Deseasonalized Sunday home delivery circulation (deseasonalized average home delivery sales on Sundays during the month)

TIME: Trend (TIME = 1 for January, 1987, TIME = 2 for February, 1987, …, TIME = 135 for March, 1998)

UEMPLOY: Unemployment rate in the state of Massachusetts

SUNHOME$: Average price of a home delivery subscription for Sunday only

HSUNRTRN: Sunday home delivery return percentage

The response variable is the **deseasonalized** Sunday home delivery circulation series. The raw circulation values were divided by the seasonal (monthly) indices to improve the quality of the fit by removing the monthly effects that were discussed in Boston Sunday Globe (A). The trend variable was included to make the price variable have a negative coefficient. Prices of newspapers, like many products, tend to increase over time. Since, as was discovered in Boston Sunday Globe (A), home delivery subscriptions have increased over time as well, the absence of this time variable in the model would result in a spurious positive correlation between subscription sales and price. The Massachusetts unemployment rate is a proxy for the strength of the regional economy. The Sunday home delivery return percentage is the percentage of home delivery papers that were returned undelivered back to the Globe. This variable is a measure of the effectiveness of the delivery system. An unacceptable performance by the delivery contractors (HSUNRTRN is high) may result in subscription cancellations.

Figure 1

Regression Analysis

```
The regression equation is
DSUNHOME = 372237 + 1145 TIME + 3144 UEMPLOY - 30328 SUNHOME$ - 176975 HSUNRTRN

116 cases used 19 cases contain missing values

Predictor        Coef        StDev          T        P
Constant       372237         5231      71.17    0.000
TIME          1144.88        56.20      20.37    0.000
UEMPLOY        3144.4        277.2      11.34    0.000
SUNHOME$       -30328         4650      -6.52    0.000
HSUNRTRN      -176975        33931      -5.22    0.000

S = 4881        R-Sq = 97.7%       R-Sq(adj) = 97.6%

Analysis of Variance

Source            DF          SS           MS          F         P
Regression         4 1.10587E+11 27646775076    1160.63     0.000
Residual Error   111  2644071922     23820468
Total            115 1.13231E+11
```

The Minitab regression output for one of the models that was calibrated for the analysis of single copy sales of the Sunday Globe is provided in **Figure 2.** The variables are defined as follows:

LOGDSNGL:	Logarithm of the deseasonalized Sunday single copy circulation (logarithm of the deseasonalized average Sunday single copy sales during the month)
LOGDHOME:	Logarithm of DSUNHOME defined above
LOGSNGL$:	Logarithm of the newsstand price
LOGUNEMP:	Logarithm of UEMPLOY defined above

All logarithms are base ten. The LOGDHOME variable was included to measure any **product substitution** (i.e., cannibalization) effects. If Sunday subscriptions increase, perhaps due to aggressive advertising or promotion, will there be a direct impact upon single copy sales?

Figure 2

Regression Analysis

```
The regression equation is
LOGDSNGL = 9.57 - 0.710 LOGDHOME - 0.406 LOGSNGL$ + 0.0947 LOGUNEMP

134 cases used 1 cases contain missing values

Predictor        Coef        StDev          T        P
Constant       9.5688       0.3231      29.62    0.000
LOGDHOME     -0.71015      0.05907     -12.02    0.000
LOGSNGL$     -0.40568      0.02972     -13.65    0.000
LOGUNEMP     0.094702     0.007784      12.17    0.000

S = 0.01070       R-Sq = 96.1%      R-Sq(adj) = 96.0%

Analysis of Variance

Source            DF          SS          MS          F        P
Regression         3     0.36870     0.12290    1073.11    0.000
Residual Error   130     0.01489     0.00011
Total            133     0.38359
```

Conceptual Questions:

Examine the regression model in Figure 1:

1. Write out the estimated regression model. What percentage of the variation of the response variable DSUNHOME is explained by the right hand side variables?

2. Does each of the right hand side variables contribute to the explanatory power of the model?

3. Does this model provide evidence that Sunday Globe home delivery is a cyclical or an anticyclical product? Is this result consistent with the answer to Question 9 of Boston Sunday Globe (A)?

Examine the logarithmic model in Figure 2:

4. Write out the estimated regression model. What percentage of the variation of the response variable LOGDSNGL is explained by the right hand side variables?

5. Does each of the right hand side variables contribute to the explanatory power of the model?

6. Does this model provide evidence that Sunday Globe single copy is a cyclical or an anticyclical product? Is this result consistent with the answer to Question 7 of Boston Sunday Globe (B)?

7. Interpret the coefficient of LOGDHOME in the context of marketing the Boston Sunday Globe.

8. Interpret the coefficient of LOGSNGL$ in the context of marketing the Boston Sunday Globe.

9. Interpret the coefficient of LOGUNEMP in the context of marketing the Boston Sunday Globe.

Discussion Questions:

1. Based upon all of the information provided in these three cases, what comments and suggestions would you present to senior management at the Globe Newspaper Company?

2. If you were to continue with this study for the benefit of Globe management, what improvements, further analyses, or additional data would you wish to pursue?

MOTOR VEHICLE FATALITIES

Statistics Topics: **Time Series**
Residual Analysis
Autocorrelation
Autoregressive Models
Step Intervention

Data File: **fatalities.xls**

The motor vehicle is a necessity for many Americans. The number of vehicles in the United States has more than doubled from 1966 to 1992, while the population has only increased 28% over the same period of time. While these statistics indicate that United States citizens have become more dependent on the automobile, other data suggest that there is a down side associated with the increase in the number of vehicles on the road. For example, motor vehicles pollute the environment with their emissions and their noise, and travel time is hampered by the increased congestion that accompanies the increased number of vehicles. Furthermore, motor vehicle related accidents are one of the leading causes of death in the United States, and an increase in the number of vehicles might suggest an increase in associated fatalities over the past several decades.

In 1974 Congress passed a law requiring speed limits on interstates to be lowered to 55mph. Groups that supported this new law claim that lowering the speed limit has saved lives. How has the number of motor vehicle fatalities changed over the past few decades? Did the lowering of the speed limit result in fewer fatalities on the roads? To examine these questions, let's look at the pattern of the number of fatalities over time. **Figure 1** is a scatter plot of the **time series** of motor vehicle fatalities between 1966 and 1990.

In general, the data appear to have a downward trend over time: as time passes (the year increases), the number of automobile fatalities decreases. Do you think the relationship is linear? Let's run a simple linear regression and examine the results. In the Minitab output below, FATAL represents the dependent variable of annual vehicle fatalities and TIME represents the years 1966 to 1990 (rescaled as 0 to 24). The resulting regression equation and standard regression output is provided in **Table 1**.

The regression indicates a significant negative linear relationship between the variables, with an R^2 of about 48%. The **Durbin-Watson** statistic of 0.83 indicates that there is positive **autocorrelation** among the residuals, because it is less than the lower Durbin-Watson critical value of 1.29 (see a Durbin-Watson table in a textbook). To further verify the assumptions for a regression model, let's look at the residual diagnostic plots in **Figure 2**.

Figure 1

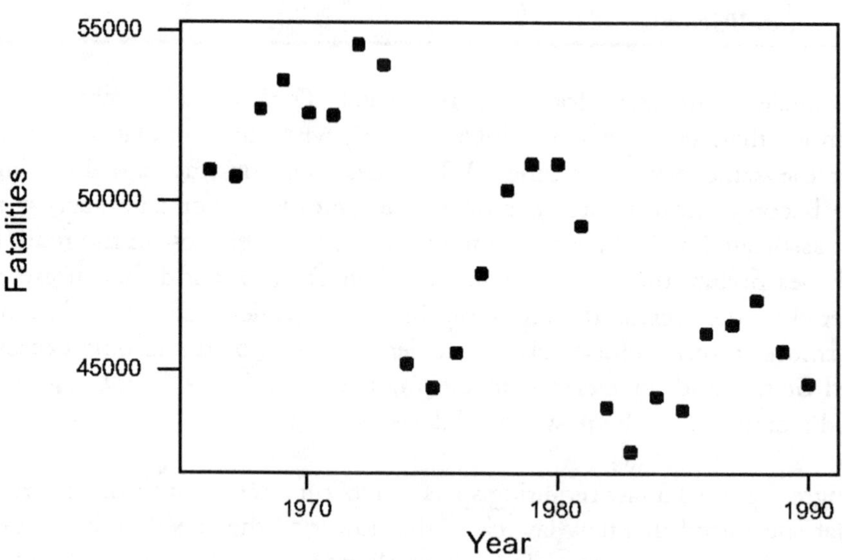

Table 1

Regression Analysis

The regression equation is
FATAL = 52703 - 355 TIME

Predictor	Coef	Stdev	t-ratio	p
Constant	52703	1074	49.06	0.000
TIME	-355.25	76.73	-4.63	0.000

s = 2766 R-sq = 48.2% R-sq(adj) = 46.0%

Analysis of Variance

SOURCE	DF	SS	MS	F	p
Regression	1	164066880	164066880	21.44	0.000
Error	23	176020848	7653080		
Total	24	340087744			

Durbin-Watson statistic = 0.83

Figure 2

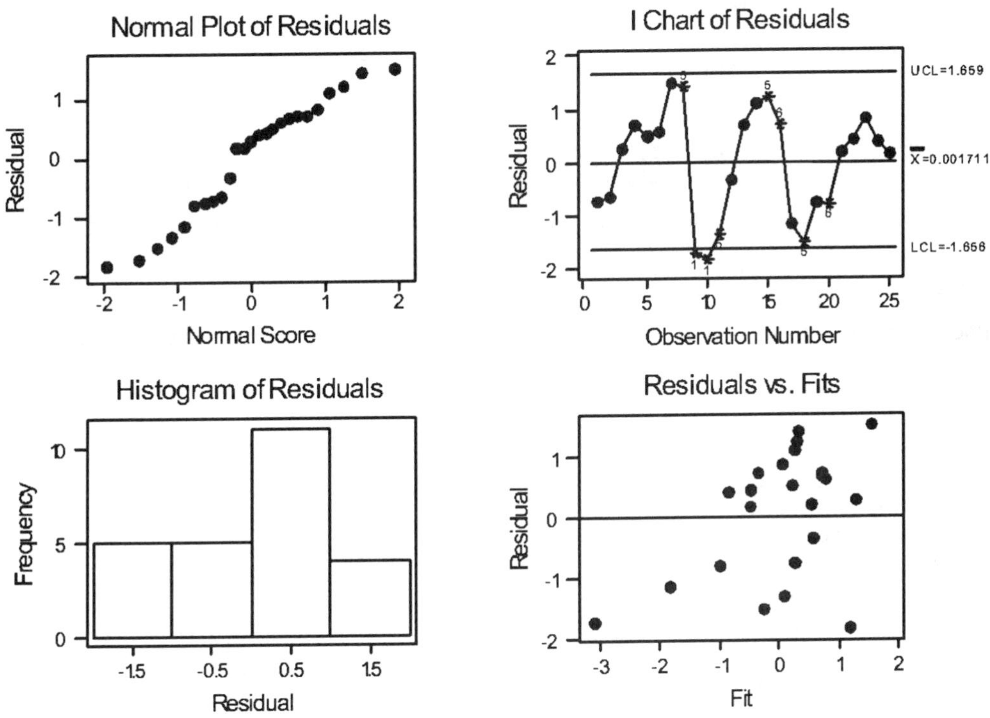

The residuals are fairly normally distributed (see "Normal Plot of Residuals" and "Histogram of Residuals" in above Figure). However, the residuals appear in a cyclical pattern in the "I Chart of Residuals," which supports the conclusion that positive autocorrelation exists between fatalities and year.

From 1966 to 1973, the relationship is what we generally expect (i.e., an increase in the number of vehicles over time leads to more fatalities). But there is a sharp decrease in the number of fatalities in 1974 – the same year the speed limit was lowered on interstates. Is this drop in vehicle fatalities significant? We ask you to examine this issue in the Analysis Questions at the end of this case.

While this drop in vehicle fatalities might suggest that the lowering of the top speed limit to 55 mph had an impact on motor vehicle fatalities, other events were taking place, which also may have had an impact on vehicular deaths. The oil crisis that the United States was experiencing during the same time period resulted in higher gasoline prices and prompted Americans to drive less, thereby leading to fewer vehicles on the road and fewer motor vehicle related fatalities. When the oil crisis ended, the number of fatalities increased each year until 1980. However, in subsequent years the number of fatalities began to decrease again before an eventual upward trend in the mid-to-late eighties. The overall relationship between fatalities and year is cyclical, suggesting that there may be a pattern to the fluctuation.

How can we model time-series data with inherent **autocorrelation**? We will improve the retrospective fit and future forecasts if we consider an **autoregressive model.** A first-order autoregressive model accounts for the association between adjacent values in a time-series and has the form:

$$FATAL_i = A_0 + A_1 FATAL_{i-1} + \delta_i,$$

where $FATAL_i$ = fatalities in year i,
$FATAL_{i-1}$ = fatalities in year i-1, and
δ_i = random error term.

The explanatory variable is also called a **lagged variable**, because it lags behind the response variable in time. Now, let's run the regression using the lagged variable.

Table 2

Regression Analysis
The regression equation is
FAT(i) = 9984 + 0.789 FAT(i-1)

24 cases used 1 cases contain missing values

Predictor	Coef	Stdev	t-ratio	p
Constant	9984	6615	1.51	0.145
FAT(i-1)	0.7892	0.1357	5.81	0.000

s = 2446 R-sq = 60.6% R-sq(adj) = 58.8%

Analysis of Variance

SOURCE	DF	SS	MS	F	p
Regression	1	202229584	202229584	33.81	0.000
Error	22	131584080	5981095		
Total	23	333813664			

Unusual Observations

Obs.	FAT(i-1)	FAT(i)	Fit	Stdev.Fit	Residual	St.Resid
9	54052	45196	52640	893	-7444	-3.27R
17	49301	43945	48891	508	-4946	-2.07R

Durbin-Watson statistic = 1.41

Note, that the relationship is stronger, with an R^2 of about 61%. More importantly, the Durbin-Watson statistic is greater as well. It is less than the critical value of 1.45 obtained from the table, which suggests that the test is inconclusive as to whether autocorrelation exists or not. Now, let's study the residual diagnostics associated with this model:

Figure 3

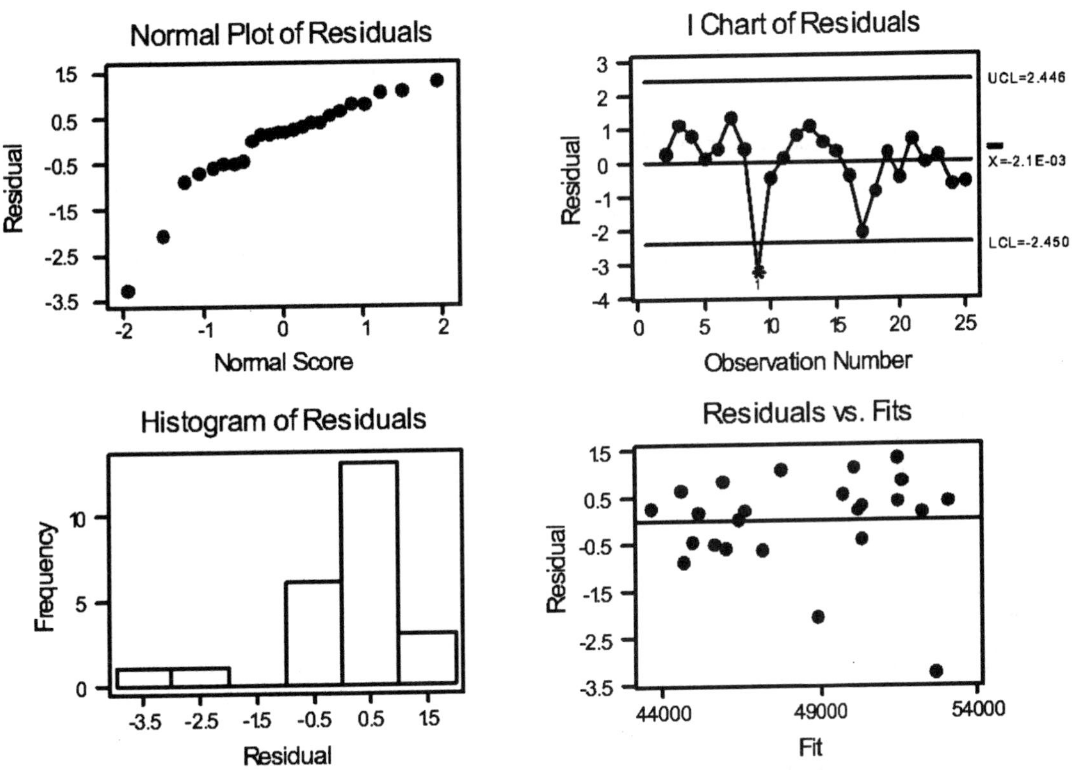

The "Normal Plot of Residuals" and the "Histogram of Residuals" indicate that the residuals are fairly symmetrical - with the clear exception of the two unusual observations. In addition, the "I Chart of Residuals" and the "Residuals vs. Fits" plots appear randomly distributed about zero, indicating no evidence of autocorrelation. The one unusual observation in the "I Chart of Residuals" represents the drop in fatalities from 1973 to 1974, which coincides with the lowering of the speed limit to 55 mph, as mentioned earlier. To account for this change in speed limit, we ask you to add an indicator variable to the regression model in the Analysis Questions at the end of this case.

Will using more prior information in our model improve our fit? If we lag the fatalities another year, we can set up a second-order autoregressive model:

$$FATAL_i = A_0 + A_1 FATAL_{i-1} + A_2 FATAL_{i-2} + \delta_i,$$

where $FATAL_i$ = fatalities in year i,
$FATAL_{i-1}$ = fatalities in year i-1,
$FATAL_{i-2}$ = fatalities in year i-2, and
δ_i = random error term.

The regression output for this second-order autoregressive model appears below.

Table 3

Regression Analysis

The regression equation is
FAT(i) = 14538 + 1.07 FAT(i-1) - 0.377 FAT(i-2)

23 cases used 2 cases contain missing values

Predictor	Coef	Stdev	t-ratio	p
Constant	14538	6915	2.10	0.048
FAT(i-2)	-0.3771	0.2101	-1.79	0.088
FAT(i-1)	1.0737	0.2088	5.14	0.000

s = 2377 R-sq = 65.5% R-sq(adj) = 62.1%

Analysis of Variance

SOURCE	DF	SS	MS	F	p
Regression	2	214840208	107420104	19.01	0.000
Error	20	113030664	5651533		
Total	22	327870880			

SOURCE	DF	SEQ SS
FAT(i-2)	1	65436084
FAT(i-1)	1	149404128

Unusual Observations

Obs.	FAT(i-2)	FAT(i)	Fit	Stdev.Fit	Residual	St.Resid
9	54589	45196	51986	951	-6790	-3.12R
10	54052	44525	42680	1778	1845	1.17 X

R denotes an obs. with a large st. resid.
X denotes an obs. whose X value gives it large influence.

Durbin-Watson statistic = 1.87

The higher p-value (.09) of the two-year lagged variable indicates that it is a weaker variable in the regression than the one-year lagged variable. However, note that the standard error of the estimate (s) is lower, the R^2 (R-sq) is higher, and the Durbin-Watson statistic (1.87) now decisively indicates the absence of autocorrelation in this second-order model. Therefore, how should we decide which model to use for forecasting motor vehicle fatalities?

Let's compare the accuracy of the forecasts for the two models for 1991 to the actual motor vehicle fatalities. To obtain a prediction for 1991, we use our two autoregressive models to compute the number of fatalities FAT(i):

$$FAT(i) = 9984 + 0.789(44{,}599) = 45{,}173$$

and

$$FAT(i) = 14538 + 1.07(44{,}599) - 0.377(45{,}582) = 45{,}074$$

These computed values differ from the actual value (41,508) of motor vehicle related fatalities that occurred in 1991 by approximately 10% and 9%, respectively. Therefore, since the MSE for the second-order model is lower and the forecast is more accurate, we conclude that we will use our second-order model for forecasting.

Conclusions

The trend of automobile fatalities over the 25 years from 1966 to 1990 can be modeled effectively with a second-order autoregressive time-series model. However, the drop in fatalities in 1974 may be significant and an additional variable may need to be included in the model to improve its use in prediction. **Figure 4** depicts the actual data with the fatalities predicted by our second-order model.

Figure 4

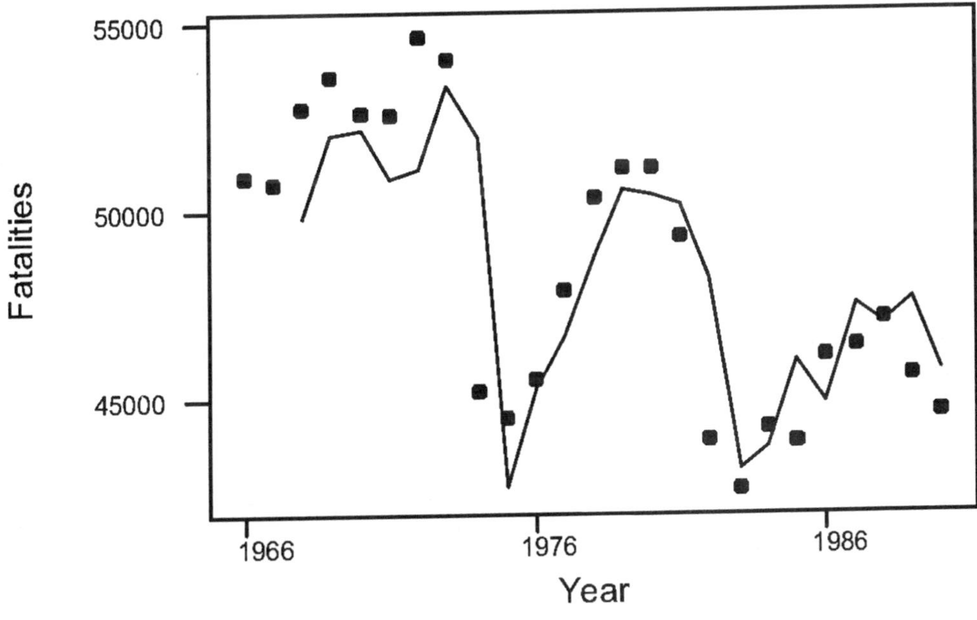

Analysis Questions:

1. Create a dummy variable to represent the change in the maximum speed limit, which we define as:

   $X_t =$ 0 for time < 1974;
   1 for time $\geq$ 1974.

   This variable is called a **step intervention**, because the graph of X_t looks like a step and the intervention is a one-time event. The coefficient of this dummy variable will represent an estimate of the size of the drop in annual motor vehicle fatalities following the change in the maximum speed limit.

2. Include this new dummy variable in the autoregressive model for fatalities. Is the fit of the model improved? Is this step-variable significant in the model? Are the regression assumptions satisfied? Is the accuracy of the forecast for fatalities in 1991 improved? Do you believe that lowering the speed limit lowered motor vehicle fatalities?

Discussion Questions:

1. In December 1995, Congress repealed a 1974 federal law that required the maximum speed limit on the interstates to be 55 mph. During 1996, in response to this act, 27 states raised their speed limits. In these 27 states motor vehicle deaths fell slightly from 24,911 in 1995 to 24,855 in 1996. However, 11 of these states reported an increase in motor vehicle fatalities, while 13 states reported a decrease, and 3 states reported no change. To include more recent data on motor vehicle deaths, how should the intervention dummy variable now be defined?

2. The total number of vehicle fatalities in 1995 and 1996 were 41,798 and 41,500, respectively. Using the model created in the Analysis portion of this case, forecast the fatalities in 1995 and 1996. How accurate are your forecasts?

Sources:

http://www.nhtsa.dat.gov/

Worsnop, Richard L. Highway Safety. In Congressional Quarterly Researcher. Washington, D.C., 1995.

QUARTERLY SALES AT THE HOME DEPOT

Statistics Topics: **Forecasting**
Exponential Smoothing
Seasonal Decomposition
Seasonal Dummy Variables
Second-Order Autoregressive Model

Data File: **homedepot.xls**

Bernie Marcus and Arthur Blank founded The Home Depot in 1978 in Atlanta Georgia. Since then it has grown to become the world's largest home improvement retailer in the Home Supply Retail Industry, with over 800 stores in the United States, Canada, and South America. The original Home Depot stores sold approximately 25,000 products, while today's stores are approximately 130,000 square feet and sell almost 50,000 products.

The mission of The Home Depot has always been to offer a wide variety of low-priced products. While there have been other cut-rate retail stores which have tried this strategy, Home Depot also recognizes the importance of customer satisfaction and hires trained sales staff, or trains them on-site. These educated sales personnel are able to provide expert advice or instructional workshops for do-it-yourself customers – which account for over 60% of the building supply industry's sales volume.

This customer-focused strategy has proven successful. After only six years in operation, Home Depot had 19 sites and reported a 118% increase in sales in 1984 ($256 million) over 1983. By 1986 Home depot was operating 50 retail stores. However, this rapid expansion took its toll – earnings dropped by 42% in 1985 – caused primarily by the impact of the rapid growth on long-term debt. A more conservative approach to growth was needed. In the late 1980's stores were opened in existing markets – as opposed to new markets – to make more efficient use of current spending, and a computerized inventory system was installed to lower inventory costs, ordering costs, and distribution costs.

By 1989 Home Depot appeared to be back on track and earnings increased 46% in 1990, with a 38% increase in sales over the previous year. This growth was a reflection of a 33% increase in the number of customer transactions, and a 4% increase in the average sale per transaction. In the early 1990's the company expanded into the northeast with 75 new stores – banking on the dense consumer markets of the region. However, as competitors also opened warehouse-type stores and the market became saturated, sales leveled off by 1993. It appeared that after experiencing tremendous growth from 1989 to 1993, The Home Depot had entered into the stable growth phase, as a result of industry saturation and increased competition from other "megastores."

This case was prepared by Bethanie Lydon, under the direction of Professor Norean Sharpe, with the assistance of Irena Vekselberg, as a basis for class discussion.

The Home Depot has continued to focus on customer satisfaction and remains confident about maintaining growth through 2000. Is this possible? Their compound growth rate for the first half of the 1990's was 35%, but can this success be repeated? As corporations and products mature, growth rates typically slow down – unless new markets can be reached.

The objective of this case is to use Home Depot's historical sales figures to forecast future quarterly sales. This is referred to as **time series** analysis, since the data are measured over time and we are trying to find a pattern in the historical data. We will use The Home Depot's quarterly sales from 1994 to 1998 to develop alternative forecasting models. We will then compare the forecasts for first quarter sales in 1999 to actual sales to determine the most appropriate forecasting model.

First, we examine the quarterly data from 1994 to 1998 and try to find any obvious patterns or trends in Home Depot's sales. The scatter plot for Home Depot's quarterly sales from 1994 to 1998 is shown in **Figure 1**.

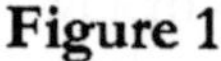

Figure 1

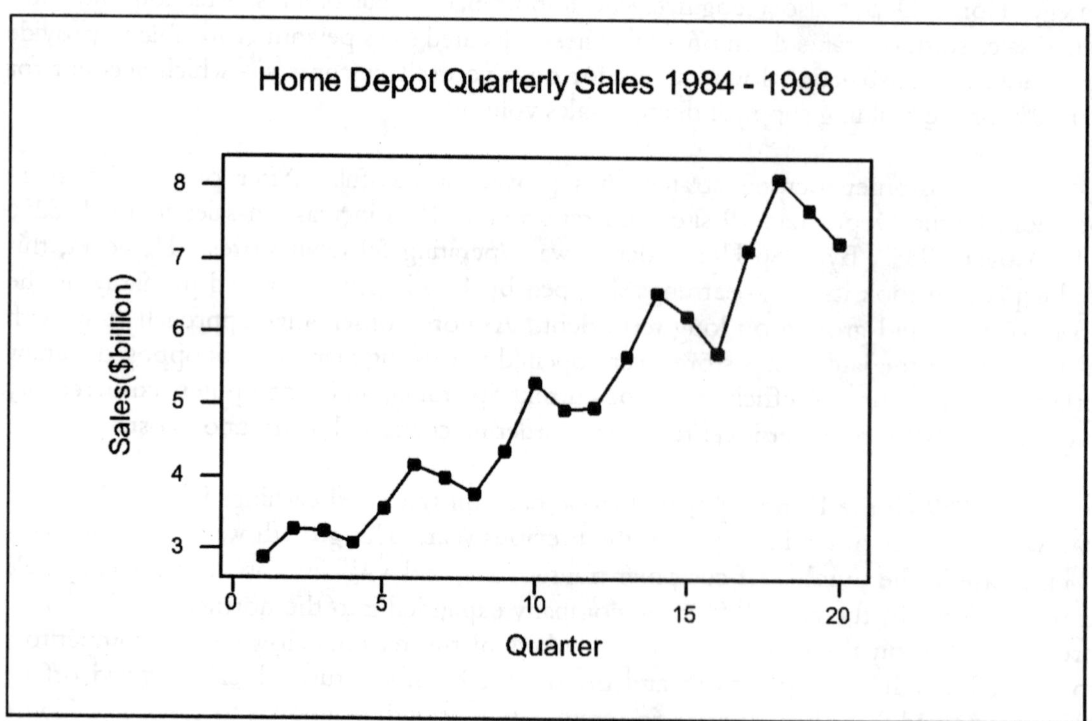

From the graph, we can see that Home Depot's quarterly sales follow a seasonal pattern with the second quarter having the highest sales and the first and fourth quarter having the lowest sales. This makes sense since builders and homeowners tend to construct and renovate homes during the warmer seasons, rather than during the colder seasons.

There is also a long-term positive trend in the data, which suggests continued growth in sales since 1994. Thus, from our visual analysis of the data, we hypothesize that Home Depot sales follow a seasonal pattern with an upward trend.

Model I

We first fit a linear regression model to the data. This model accounts for the upward trend in the data, but not the seasonal component. The output for this model in Excel is shown in **Table 1**, and for Minitab is shown in **Table 2.** From the output in both Table 1 and Table 2, we can see that the regression model is significant, with the p-value less than .001 and a high R^2 (92%). The regression equation is:

$$\text{Sales} = 2.29 + 0.267 \text{ Quarter},$$

which means that Home Depot's sales increase by approximately $.267 billion per quarter. To get our forecast for the first quarter of 1999, we put the quarter number into the equation. Since the quarters for 1994 to 1998 were labeled 1 to 20, the first quarter for 1999 is quarter 21. The result is a forecast of $7.897 billion for the first quarter of 1999.

Table 1

SUMMARY OUTPUT

Regression Statistics	
Multiple R	0.961
R Square	**0.923**
Adjusted R Square	0.919
Standard Error	0.470
Observations	20.000

ANOVA

	df	*SS*	*MS*	*F*	*Significance F*
Regression	1.000	47.541	47.541	215.569	0.000
Residual	18.000	3.970	0.221		
Total	19.000	51.511			

	Coefficients	*Standard Error*	*t Stat*	***P-value***	*Lower 95%*	*Upper 95%*	*Lower 95.0%*	*Upper 95.0%*
Intercept	**2.285**	0.218	10.477	**0.000**	1.827	2.744	1.827	2.744
Quarter	**0.267**	0.018	14.682	**0.000**	0.229	0.306	0.229	0.306

Table 2

Regression Analysis

The regression equation is
Sales = **2.29** + **0.267** Quarter

Predictor	Coef	StDev	T	P
Constant	2.2856	0.2181	10.48	0.000
Quarter	0.26737	0.01821	14.68	**0.000**

S = 0.4696 **R-Sq = 92.3%** R-Sq(adj) = 91.9%

Analysis of Variance

Source	DF	SS	MS	F	**P**
Regression	1	47.540	47.540	215.59	**0.000**
Residual Error	18	3.969	0.221		
Total	19	51.509			

Unusual Observations

Obs	C18	Sales2	Fit	StDev Fit	Residual	St Resid
18	18.0	8.139	7.098	0.172	1.041	2.38R

R denotes an observation with a large standardized residual

The linear model and residual plot appear below as Figure 2 and Figure 3, respectively (both in Minitab). The residual plot is a plot of the residuals over time (versus quarter). From the graphs, it is evident that there is seasonality in the data that has not been accounted for in this model and the magnitude of the residuals is growing over time.

Figure 2

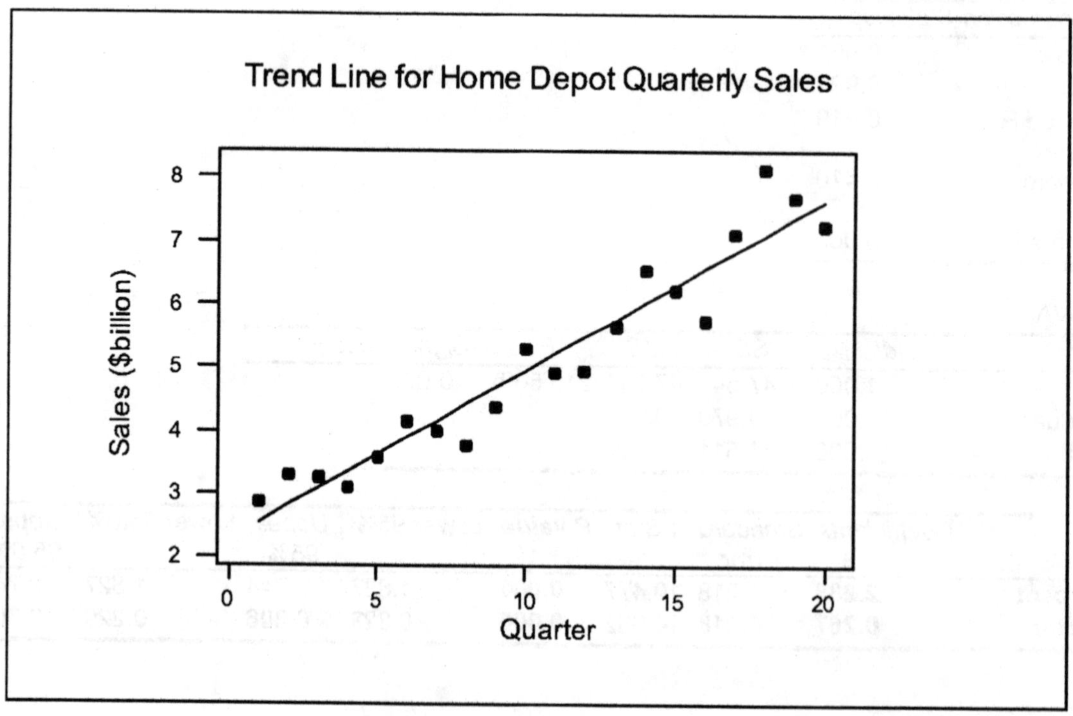

Figure 3

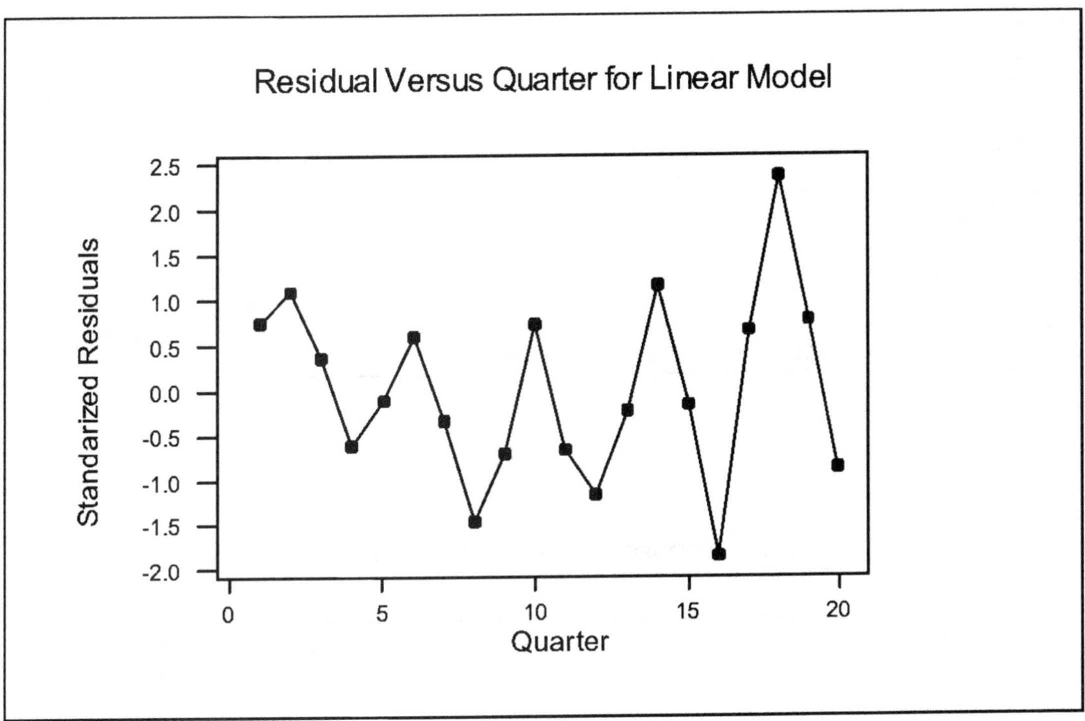

Model II

Next, we attempt to fit a single exponential smoothing (SES) model to the data. Exponential smoothing uses all of the historical data in the time series to generate forecasts, but places more weight on the more recent values. It is usually used for a stationary time series, and its forecasts are flat. This SES model does not account explicitly for seasonality, and uses a weight (smoothing coefficient) to adjust emphasis on prior values. The forecasts for exponential smoothing are computed as follows:

$$\hat{Y}_{t+1} = \hat{Y}_t + \alpha(Y_t - \hat{Y}_t),$$

where α = the smoothing coefficient. The output from this model in Minitab is shown in **Table 3**, and the graph with the smoothed data is shown in **Figure 4**. Note that $\alpha = 1.812$.

Table 3

```
Single Exponential Smoothing
Data          Sales
Length        20.0000
NMissing      0

Smoothing Constant
Alpha: 1.81213

Accuracy Measures
MAPE: 7.79278
MAD:   0.39693
MSD:   0.25701

 Row   Period  Forecast      Lower      Upper
   1       21   7.22470    6.25222    8.19718
```

Figure 4

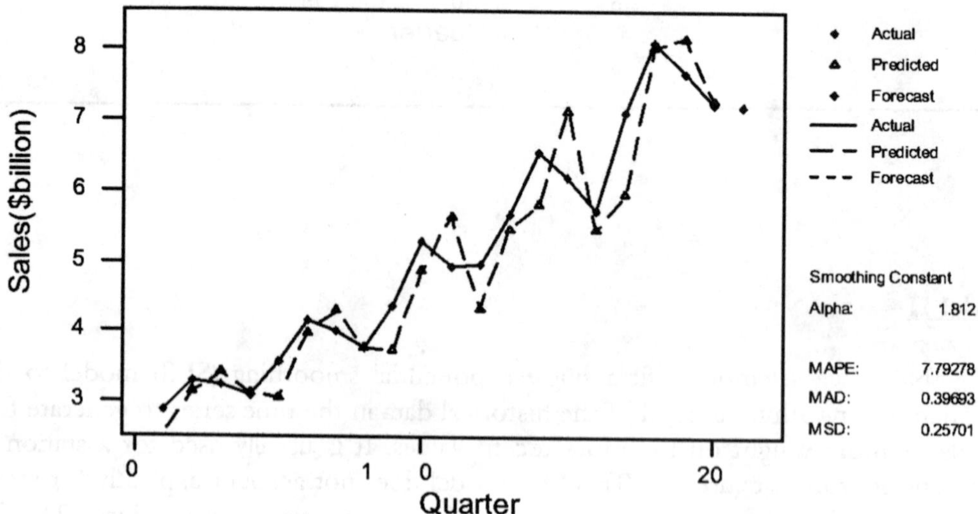

Model III:

Next, let's examine the Seasonal Decomposition model with a linear trend. The Minitab output for this model is shown in **Table 4** and the graph of the predicted values for this model is shown in **Figure 5.**

Table 4

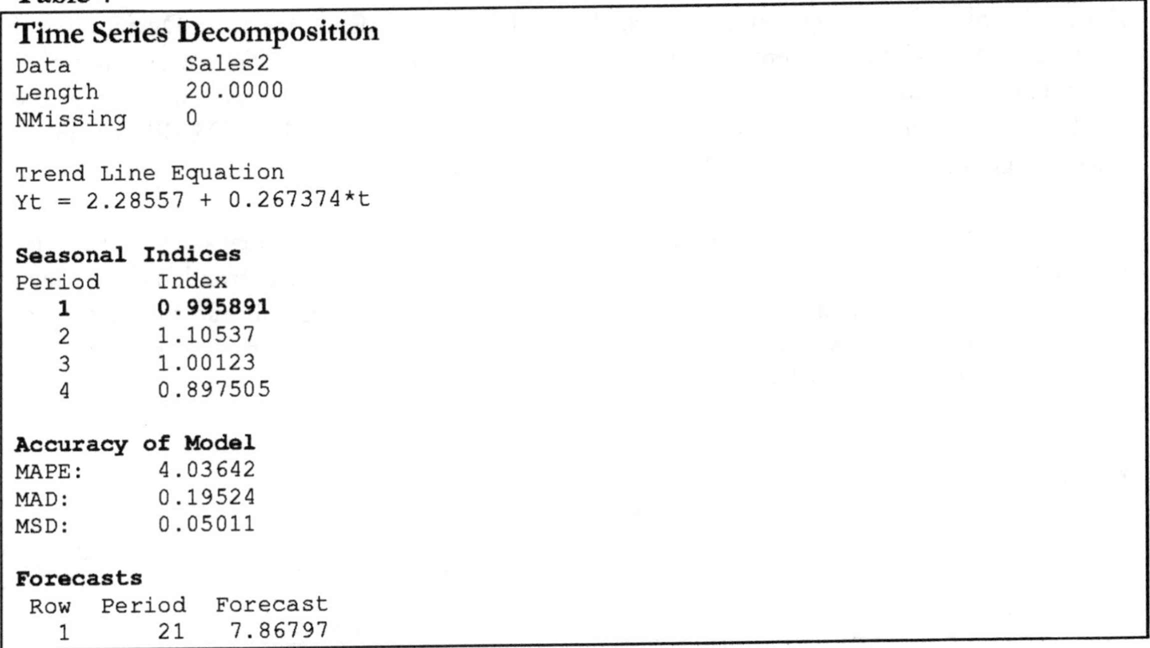

Time Series Decomposition

```
Data          Sales2
Length        20.0000
NMissing      0

Trend Line Equation
Yt = 2.28557 + 0.267374*t
```

Seasonal Indices

Period	Index
1	**0.995891**
2	1.10537
3	1.00123
4	0.897505

Accuracy of Model

MAPE:	4.03642
MAD:	0.19524
MSD:	0.05011

Forecasts

Row	Period	Forecast
1	21	7.86797

Figure 5

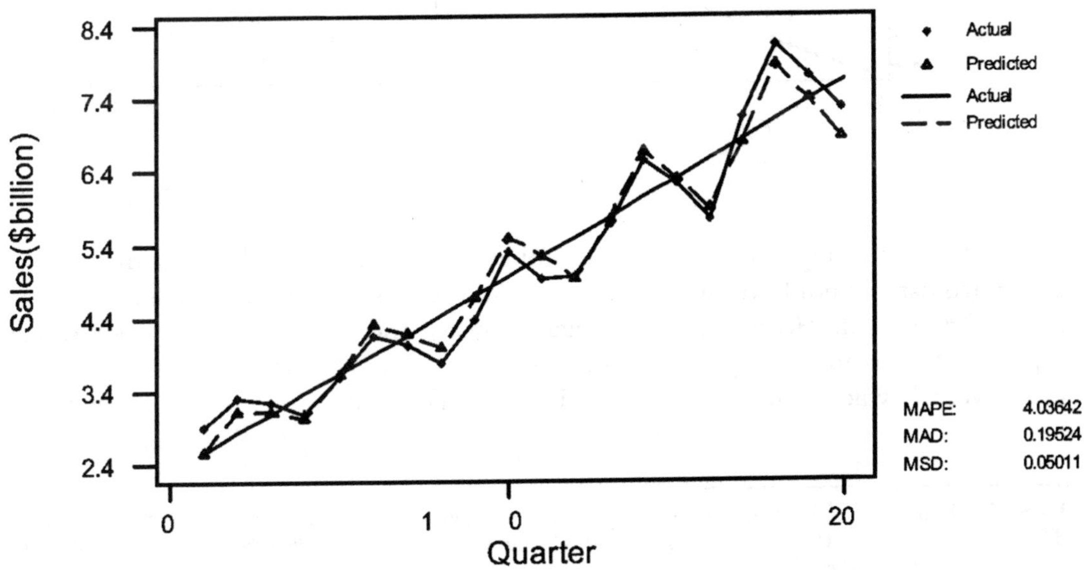

From the graph and the statistical output, we can see that the model has a MAPE (Mean Absolute Percentage Error)[1] of 4.04% and a MAD (Mean Absolute Deviation)[2] of $195.24 million. The measurements of the magnitude of the residual errors are only useful in a comparison with the same measures of the other models. However, we can say that the MAD is better than that for the SES model ($396.93 million) and the MAPE of 4% is relatively good, compared to the MAPE of 7.8% for SES.

Another way to assess a model's accuracy is to examine the residual analysis. In order to evaluate the Seasonal Decomposition model, we should look at the detrended data, which should have no trend component; the seasonally adjusted data, which should follow a linear trend; and the seasonally adjusted and detrended data, which should have no pattern. The Component Analysis for Sales is shown in **Figure 6**.

Figure 6

Component Analysis for Sales

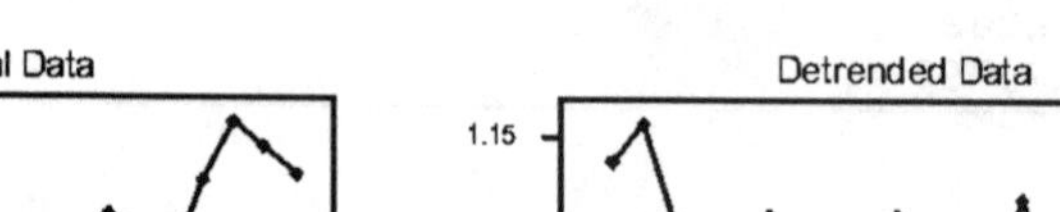

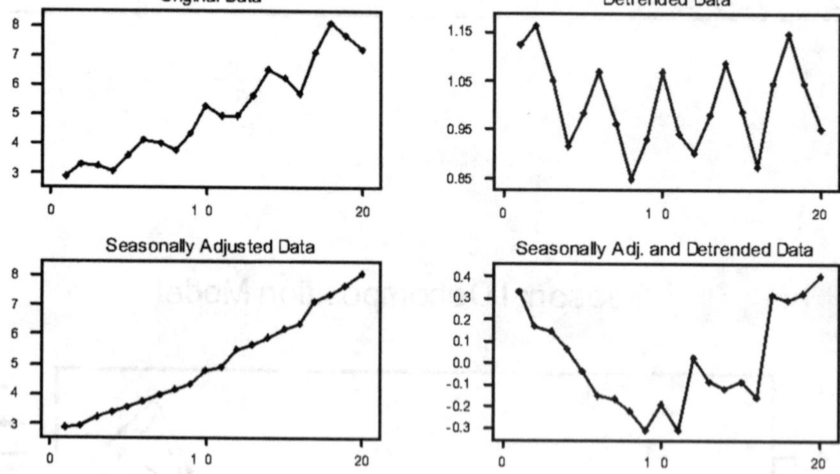

From the output above we can examine how well the model fits the data. The detrended data do not have an upward trend and the seasonally adjusted data do not have seasonality present. However, the seasonally adjusted and detrended data, which should represent the irregular component of the time series, does appear to have a pattern, most likely cyclical, which is not accounted for in the model. Below in Table 5 are the forecasts

---

[1] MAPE is the measure of the relative or percentage error, which can be calculated by taking a sum of absolute values of the difference between the actual and a fitted value divided by the actual, and then dividing the sum by the number of observations and multiplying by 100.

[2] MAD is the measure of the average of the absolute discrepancies between the actual and fitted values. It can be calculated by taking a sum of absolute differences between the actual and a fitted value and dividing the sum by the number of observations.

for the first quarter sales in 1999 using the three different approaches shown here in the case.

Table 5

Home Depot First Quarter Sales For 1999 (billions)	Forecast	Actual	Absolute Difference	% Difference
Linear Regression Model	**$7.897**	$8.952	1.055	11.79%
Single Exponential Smoothing Model	$7.225	$8.952	1.727	19.29%
Seasonal Decomposition	**$7.868**	$8.952	1.084	12.11%

Conceptual Questions:

1. Examine the output in Table 4. How are the fitted values from the trend line and the seasonal indices used to obtain the seasonal forecasts using this approach to modeling seasonal data? What do the seasonal indices represent? What information does the seasonal index for the second quarter give us?

Analysis Questions:

1. Compare the strengths and weaknesses of each of the models developed in the case.

2. Develop a first- and second-order autoregressive model and forecast the first quarter sales for 1999. How does this forecast compare to those developed in the case and to the actual sales?

3. Create seasonal dummy variables (i.e., Q1= 1 if first quarter, Q2= 1 if second quarter, and Q3= 1 if third quarter). Add these dummy variables to the linear regression model used in the case (which already includes the trend variable) and generate a forecast for the first quarter sales of 1999. Compare to other forecasts.

4. Use logarithms to transform the sales variable then reapply the seasonal model developed in question 3. Now, how do you interpret the coefficients of each variable in the model? Generate a forecast for 1999 first quarter sales and compare to others.

Discussion Questions:

1. Which model would you recommend to the sales managers at The Home Depot for generating sales forecasts, which they will then use to set sales targets, quotas, and commissions? Explain your reasoning.

Sources:

http://www.homedepot.com

Grossman, W.R. (1997). The Home Depot, Inc. In Pederson, J.P. (Ed.) International Directory of Company Histories (pp. 238-240). Detroit, MI: St. James Press.